LESLIE M. HOCKING

Department of Mathematics, University College London

Optimal Control

An Introduction to the Theory with Applications

CLARENDON PRESS · OXFORD

1991

Oxford University Press, Walton Street, Oxford OX2 6DP

Oxford New York Toronto
Delhi Bombay Calcutta Madras Karachi
Petaling Jaya Singapore Hong Kong Tokyo
Nairobi Dar es Salaam Cape Town
Melbourne Auckland

and associated companies in
Berlin Ibadan

Oxford is a trademark of Oxford University Press

Published in the United States
by Oxford University Press, New York

British Library Cataloguing in Publication Data

Hocking, Leslie M.
Optimal control.
1. Optimal control theory
I. Title
629.8312

ISBN 0–19–859675–8

Library of Congress Cataloging in Publication Data

Hocking, Leslie M.
Optimal control: an introduction to the theory with applications
Leslie M. Hocking.
p. cm.—(Oxford applied mathematics and computing science series)
Includes bibliographical references and index.
1. Control theory. 2. Mathematical optimization. I. Title.
II. Series.
QA402.3.H63 1991 629.8'312—dc20 90–44894

ISBN 0–19–859675–8 (hardback)
ISBN 0–19–859682–0 (pbk.)

Set by APS Ltd. Salisbury, Wilts.
Printed in Great Britain by
Bookcraft (Bath) Ltd.
Midsomer Norton, Avon

Preface

Optimal control theory has grown rapidly in the last 30 years in three ways. Rigorous mathematical analysis has put the subject on a sound theoretical basis. Constructive methods have been developed to allow solutions to well-posed problems to be found. The mathematical modelling of optimal control problems arising in a wide variety of contexts has been achieved. As well as the inherent interest of the topic to the more theoretically minded researchers, it is recognized as an important tool for the solution of problems that occur naturally in such diverse fields as medicine, dynamics, ecology, economics, oil recovery, and electric power production.

The essential properties of an optimal control problem are that we have a system which evolves in time according to certain laws. These laws are embodied in equations that contain elements which can be adjusted from outside the system, known as the controls. By suitable choice of these controls it may be possible to force the system into a desired target state. If this can be done at all, it can usually be done in many different ways. The choice among the successful controls of the optimal ones is governed by the necessity to make some quantity, known as the cost, as small as possible. A simple example would be that of controlling a population by means of a programme of culling or importation of predators. The target would be a population of a certain size, and the cost might be based on the numbers of predators introduced or destroyed, or on the time taken to reach the target, or a combination of both ingredients.

There are two main branches of optimal control theory, depending on whether the equations governing the evolution of the system are deterministic or stochastic. Only the former type will be discussed in this book. The development of the subject stems largely from the work of the Soviet mathematician Pontryagin and his colleagues, and the maximum principle that bears his name, familiarized to many by the publication in English of his group's textbook in 1962, is at the heart of the material presented in this book. Many books on optimal control theory of varying levels of sophistication have been published since then, so the natural questions to ask are: why has the author written yet another textbook, and how has he managed to persuade the publishers to produce it? There are two parts to my answer to the first question.

In the first place, it seemed to me that the subject needed an introductory text for those approaching the topic for the first time and without sophisticated mathematical analysis at their disposal. Consequently I have endeavoured to present the mathematics in as lucid a way as possible, without letting notational complexity obscure the central points in the argument. Thus the proofs presented in this book are not written with complete mathematical rigour, and there are imprecisions in the arguments that would need to be removed if completely satisfactory proofs were to be given. On the other hand, I did not want to present merely a set of recipes that would enable problems to be solved, without requiring any understanding of the basis on which these methods relied for their validity. In other words, I have tried to write a book which is in the worthy tradition of Applied Mathematics of the British type. One difficulty facing any author is to decide on what preliminary mathematical knowledge the readership can be assumed to possess. When in doubt I have included a discussion of what tools I am using, on the grounds that bored readers can easily skip forward to material that they do not yet know, while it is more difficult for ignorant readers to fill the gaps in their knowledge while reading the book. I have assumed a knowledge of simple differential equations and methods for their solution, of matrices and vectors, and of the elementary properties of convex sets.

The second reason for writing this book is that it gives more attention to the applications of the theory than books at a comparable level provide. Of course, most texts do include applications of the theory to specific problems, but I have tried to include a more detailed discussion of them. I have indicated how the governing equations are derived from the statement of the problem and, when it has been solved, I have drawn attention to the consequences of the solution for the system under consideration.

The plan of the book is as follows. In Chapter 1 the salient features of the general type of optimal control problems to be discussed in the book are described; one simple elementary example is solved and two others are described here and solved later. Chapter 2 contains a summary of some of the mathematical knowledge that will be used in the remainder of the book, covering the essential points concerning vectors, matrices, and convex sets, as well as a more extensive discussion of the solution of systems of linear differential equations. From this introductory material we can proceed in one of two ways. In Part A we study the particular case of time-optimal control problems, that is, when the objective is to reach the target as quickly as possible. Chapter 3 is concerned with the concept of controllability, which is a necessary precursor to any study of optimal control. How can we find the best control when the set of successful controls is empty? For linear problems (to which most of the discussion in Part A is confined) specific criteria for controllability are obtained. Chapter 4 contains the constructive method for the solution of linear time-optimal problems and the identification of non-uniqueness of the optimal control. All these chapters contain a discussion of specific applications and Chapter 5 presents a more extensive treatment of further problems. Part B is concerned with the general problem, linear or nonlinear, with a wide class of cost functions, to which the Pontryagin maximum principle applies. This principle is explained in its basic form in Chapter 6 and applied to some simple problems. Chapter 7 sees some extensions of the theory to cover partially restricted targets, terminal costs, and non-autonomous problems. In Chapter 8 there is a detailed discussion of an important subclass of problems—linear state equations with quadratic cost functions. These chapters also contain a number of examples of the application of the theory. Chapter 9 contains an outline of the proof of the Pontryagin maximum principle. Applications of a greater degree of difficulty are discussed in Chapter 10, some of which introduce new extensions to the theory or to the techniques employed in their solution. Part C consists of Chapter 11, which contains detailed discussions of four problems to which the ideas of optimal control theory have been successfully applied, and Chapter 12, which deals with numerical methods suitable for solving optimal control problems. At the end of each chapter there is a set of exercises which are not just routine applications of the techniques described in the appropriate chapter, but contain new applications or involve some new principles. The book also contains an annotated bibliography, outline solutions to the exercises, and an index.

For many years I have given a course of lectures to final-year undergraduate students in the Mathematics Department at University College London. This book has developed from the notes I have prepared over the years for these lectures. Because the topic contains some novel ways in which mathematics can be applied, together with a significant amount of mathematical theory, it has proved a popular choice among the undergraduates. One of its attractions is that specific control strategies can be identified quite readily and with a minimum of manipulative algebra for problems for which the optimal choice of control is far from obvious. The book will also, I hope, be useful for other scientists, engineers, or economists who need an introduction to the theory of optimal control because of its applicability to the real problems that they encounter.

In conclusion, I should like to thank the members of course C361. Their reaction to the material presented to them in my lectures has helped in the modification and improvement of what I have tried to teach them, and so has influenced the contents of this book. I am also indebted to Dr G. R. Walsh and an anonymous referee who have helped to remove some, if not all, of my mistakes.

London L. M. H.
1990

To the reader

Chapters 1 and 2 are introductory. Either Part A or Part B can be read first, as they are self-contained. The abbreviations TOP for time-optimal principle and PMP for Pontryagin maximum principle are used in Parts A and B respectively. If you wish to learn how to solve optimal control problems without being concerned with why the methods work, a possible course would be to read Chapters 6, 7, 8, and Chapter 10 (Sections 1–3). This could be followed by Chapter 9 (Section 1), Chapter 4 (Section 4), and Chapter 5, in which TOP is deduced from PMP and applied to time-optimal problems.

Sections of chapters are referred to, for example, as §4.1. In some sections, numbered problems are discussed and these are referenced, for example, as §4.4(2), this being Problem 2 in Section 4 of Chapter 4. Numbered results within a section are referenced similarly. Equations, Figures, and Exercises are numbered consecutively in each chapter with the chapter number included. For example, equation number 22 of Chapter 4 is referred to as (4.22), the third figure in Chapter 6 as Fig. 6.3, and the fourth exercise in Chapter 5 as Exercise 5.4.

Bold type is used for vectors and matrices. A subscript is used for a component of a vector, a superscript 0 for the value at time t_0 or 0, and a superscript 1 for the value at time t_1. A superscript equal to two or more will usually denote the appropriate power. Thus x_3^1 is the value of the third component of the vector $\mathbf{x}$ at time t_1, and $x_3^2 = (x_3(t))^2$. A dot above a scalar, vector or matrix denotes the time derivative. The scalar product of two vectors is denoted by $\mathbf{x} \cdot \mathbf{y}$ and also by $\mathbf{x}^{\mathrm{T}}\mathbf{y}$, where $\mathbf{x}^{\mathrm{T}}$ is a row vector, the transpose of the column vector $\mathbf{x}$.

Sets of points or functions are denoted by script type, for example $\mathscr{T}$ or $\mathscr{U}$. The symbols ∂ and Int denote the boundary and the interior of a set, respectively. The Euclidean space of n-dimensional real vectors is denoted by $\mathscr{R}^n$.

Contents

1 Optimal control problems

Though foil'd he does the best he can.

John Dryden

We begin by describing optimal control problems and give formal statements of the types of problem which we shall be considering.

1.1. Introduction

All vehicles require controls if they are to function in a desired manner. These controls may, for example, steer the vehicle, accelerate it, or stop it. We can therefore think of the vehicle in two parts: the engine, body, and seats, and the controls: accelerator, brake and steering wheel. The controls may be activated by a driver in the vehicle or remotely by radio signals from outside. It is also possible for the controls to be worked automatically by a feedback device. There are two ways in which controls may be needed by a moving system. They are needed to move the system from an original state to a desired one, but they may also be needed for continuous operation in order to counter deviations from a desired path. For example, a ship may be set on a straight course, but without constant applications of touches on the rudder it will tend to wander from the desired bearing. This is because the system is unstable and unavoidable minute disturbances will produce a large drift from the correct course unless controls are applied.

It is not only in the mechanics of motion that control problems can exist. Other examples occur in the control of growth processes in organisms and populations. Here the controls may take the form of added nutrients or pesticides. Medical examples are the control of a malfunction by the administration of a drug. In economics, the operation of a company is dependent on financial controls, and the manager of a factory has some control over the rate and quality of production.

In controlled systems of this type, it is generally possible to steer the system from one state to another by application of the controlling devices in many ways. In order to choose between the various strategies, we can attach some cost to their operation, such as the time taken to reach the goal, the expenditure on fuel, or the amount of energy utilized. The decision that has then to be made is to choose from all the successful strategies those that can be performed with minimum cost. This is the essence of the *optimal control problem.* There is, of course, no guarantee that the optimal control will be unique or that it will exist. Non-uniqueness may be removed by the addition of a further ingredient to the cost which will discriminate between the rival candidates for optimality. Non-existence usually occurs when the mathematically minimum cost is not achievable, but costs arbitrarily close to the minimum can be achieved. Again, the addition of an extra component to the cost may make the minimum reachable.

1.2. Plant growth

To illustrate the salient features of the type of optimal control problem that this book is concerned with, let us look at an example of a very simple kind, for which the solution can be found by elementary methods. Suppose a market gardener has a number of plants that he or she wishes to grow to a certain height by a given date. Their natural rate of growth can be accelerated by artificial lighting to reduce the hours of darkness when no growth takes place. After choosing appropriate units, we can model this process by a differential equation for the height $x(t)$ of the plant at time t, of the form

$$\frac{\mathrm{d}x}{\mathrm{d}t} = 1 + u, \tag{1.1}$$

where $u(t)$ measures the excess rate of growth produced by the artificial light. We may suppose that the height is zero at the start, but that a height of two units is required at the end, after one time unit has elapsed. Thus we require

$$x(0) = 0, \quad x(1) = 2. \tag{1.2}$$

The extra growth produced by keeping the light burning will incur a financial cost, depending on the size of the control, and a possible cost function has the form

$$J = \int_0^1 \tfrac{1}{2} u^2 \, dt. \tag{1.3}$$

The objective is to find the control variable u that will produce a solution of (1.1) satisfying (1.2) and at the same time allow J to take its minimum value.

The solution of (1.1) that satisfies the first condition of (1.2) is

$$x(t) = \int_0^t [1 + u(\tau)] \, d\tau, \tag{1.4}$$

and when we apply the second condition of (1.2) we see that $u(t)$ must satisfy the condition

$$\int_0^1 u(\tau) \, d\tau = 1. \tag{1.5}$$

This condition is all that is needed to determine the successful control strategies. In order to find the optimal one we must consider the cost function J. From (1.3) we have

$$\begin{aligned} J &= \int_0^1 \tfrac{1}{2} [(u-1)^2 + 2u - 1] \, dt \\ &= \int_0^1 \tfrac{1}{2} (u-1)^2 \, dt + \int_0^1 u \, dt - \tfrac{1}{2} \\ &= \int_0^1 \tfrac{1}{2} (u-1)^2 \, dt + \tfrac{1}{2}, \end{aligned} \tag{1.6}$$

using (1.5). The minimum value of J is then clearly equal to $\frac{1}{2}$, since the integral is non-negative. Moreover, this minimum is achieved by letting $u(t)$ take the value 1 for all t between 0 and 1. Thus the optimal cost is $\frac{1}{2}$, the optimal control is $u(t) = 1$, and the optimal trajectory, that is, the solution of (1.1) using the optimal control, is $x(t) = 2t$. It is only very rarely that an optimal control problem can be solved by such elementary methods. It is the purpose of this book to describe general methods for the solution of optimal control problems that do not depend on special methods which are only useful in individual cases.

Before giving the general formulation of an optimal control problem, we can introduce some of the nomenclature by reference to this special example. The state of the system is described by the single variable x, which satisfies the state equation (1.1), which includes a single control variable u. The initial and final states of the system are given by (1.2); the final state is also called the target and, in this case, the target is fixed and is to be reached in a fixed time. There is no bound to the size of the control that can be employed, but there is a cost function (1.3) that depends on the square of the control variable.

1.3. General formulation

The general statement of optimal control problems of the type we are going to consider can now be given. We are interested in the behaviour of some system which is evolving according to some deterministic laws. In other words, we are *not* going to consider systems whose behaviour is random or stochastic. The system has features that can be quantified by the values of many variables, but we suppose that the relevant behaviour of the system requires n variables to characterize it. The identification of these variables and the description of the system in terms of them is the major task of constructing a mathematical model of the system, but this aspect will not be treated here. We suppose that this task has been completed and the whole behaviour of the system is fixed by the values of the *state vector* $\mathbf{x}$, whose components x_i, $i = 1, 2, \ldots, n$, are the *state variables.* The system evolves with time t, so that the state variables are functions of t, and they are governed by n first-order differential equations, the *state equations*, which have the general form

$$\dot{\mathbf{x}} = \mathbf{f}(t, \mathbf{x}, \mathbf{u}), \tag{1.7}$$

where the dot denotes a time derivative and $\mathbf{f}$ has n components f_i. The state equations depend on the state of the system and on the time, and they also depend on the *control variables* u_i, $i = 1, 2, \ldots, m$, which form the m-dimensional *control vector* $\mathbf{u}$. We shall assume that these control variables are integrable functions of t. It would simplify the mathematics considerably if we could strengthen this assumption and require them to be continous but, as we shall see, that would be too restrictive. The admissible control variables will be said to belong to the set $\mathcal{U}$; in addition to the requirement already mentioned, we may, in some problems, also restrict the maximum and minimum values that any of the control variables can take. For instance, we may want to insist that u_i lies between a_i and b_i. In that case, we can make a linear change in the variable and replace u_i by another variable v_i, defined by

$$u_i = \tfrac{1}{2}(a_i + b_i) + \tfrac{1}{2}(a_i - b_i)v_i, \tag{1.8}$$

so that v_i is also integrable and

$$-1 \leq v_i(t) \leq +1 \tag{1.9}$$

for all values of the time t. Hence the set $\mathscr{U}$ of all the admissible controls will consist of integrable functions of t which are either without bounds (although, of course, any particular control will be a bounded function of t) or bounded by the extreme values -1 and $+1$. If the controls are bounded, there is an important subset of $\mathscr{U}$ in which the control variables only take their extreme values. Any change in the value of the control is then necessarily sudden as it switches from one extreme value to the other. Such controls are said to be *bang–bang* and we shall see later that they play an important role in many problems.

The system starts in a certain configuration at an initial value of t. Thus we suppose that

$$\mathbf{x}(t_0) = \mathbf{x}^0, \tag{1.10}$$

where the initial state $\mathbf{x}^0$ and the initial time t_0 are given. Similarly, the final state is given by

$$\mathbf{x}(t_1) = \mathbf{x}^1. \tag{1.11}$$

We can distinguish different cases. The final or terminal time t_1 may be *fixed* or *free*. The terminal state may be given, or we may only require that it should lie on some curve or surface. For example, we might require that $x_1(t_1)$ should have a given value, but the other components of $\mathbf{x}(t_1)$ be without restriction. In general, the terminal condition on the state is that $\mathbf{x}^1$ must belong to some target set $\mathscr{T}$.

The final ingredient of the general statement of the optimal control problem concerns the *cost*. The most general form for the cost that we shall consider has the form

$$J = F(\mathbf{x}^1) + \int_{t_0}^{t_1} f_0(t, \mathbf{x}, \mathbf{u})\,\mathrm{d}t. \tag{1.12}$$

This cost has two parts. The first is called a *terminal cost*, as it exacts a penalty according to the terminal state of the system. Of course, it only has an effect when the terminal state is not fixed *ab initio*. The second part depends on the state of the system along the trajectory of the solution and on the time and, more importantly, on the values of the controls employed in the solution. An important special case is when $f_0 \equiv 1$ and there is no terminal cost; J is then equal to the time taken by the system to move from its initial to its final state. We then have a *time-optimal* control problem; it must, of course, be a free-time problem.

This completes the list of ingredients of the optimal control problem and the question to be answered can be formulated as follows. The state of the system can be steered from its initial to its final form by the application of certain controls belonging to $\mathscr{U}$, the evolution of the state being governed by the state equations (1.7). If there are *no* successful controls, we say that the system is *not controllable* from the given initial state to the target set $\mathscr{T}$, and there is nothing more to be said. If the system *is* controllable there will, in general, be many successful controls and to each of them there will correspond a value of the cost function J, determined by (1.12). The optimal control problem is to determine the minimum value of J, the optimal cost, and the value of the control vector $\mathbf{u}$ for which this cost is obtained, which is the optimal control. We can also determine the corresponding solution of the state equations, which gives the optimal trajectory, as well as the final value of the time, the optimal time, and the final state, if these are not prescribed.

We have already identified some important special cases of optimal control problems. For convenience, we can classify the problems we shall encounter by the following categories. The state equations may not depend explicitly on the time, in which case we have an *autonomous* problem; if t is present in (1.7) we have a *non-autonomous* problem. The set $\mathscr{U}$ of admissible controls may be *without bounds*, *bounded*, or *bang–bang*. The final time may be *fixed* or *free*. The target may be *fixed*, *fully free*, or *partially free*, depending on whether $\mathscr{T}$ is a single point in $\mathscr{R}^n$, the whole of $\mathscr{R}^n$, or some given subset of $\mathscr{R}^n$. The cost function J may be with or without a *terminal cost* F. A special case of the integrand in the cost integral is when $f_0 \equiv 1$, which gives the *time-optimal* problem. Another special case for a single control variable u_1 (that is, when $m = 1$) is when f_0 is proportional to $|u_1|$, which is called a *fuel cost*, since the operation of the control then contributes to the cost irrespective of its sign. A third example is when f_0 is proportional to u_1^2; with this *quadratic cost* some progress towards a general solution can be made.

To illustrate the foregoing description of the general problem, in the next section a particular problem will be posed and described in the terminology we have developed. Although we shall not be able at this stage to derive the solution, we can make a plausible guess (which we shall later prove to be correct) at the optimal control strategy and so determine the optimal solution.

1.4. The positioning problem

In the process of manufacturing microcircuits on wafers of silicon, it may be necessary to move a minute object to a designated position. Because of the fragility of the material and the need to avoid contamination, it is advisable that this movement be accomplished without direct contact between the wafer and the operator's fingers or tools. Possible means by which the movement can be achieved are directed air jets and electric or magnetic fields. Consider the one-dimensional problem of moving the object along a line, with a coordinate x_1 measuring the displacement of the object from its desired position. The state of the system is determined by the position and velocity of the object, so that we need another state variable x_2 for the velocity. The force acting on the object is controllable and we can quantify it by a control variable u_1. We can suppose that initially the object is at distance X from home and is at rest. We wish to move it to the home position and for it to remain there, so the final state is given by $x_1 = 0$, $x_2 = 0$. For the cost function we may suppose that we wish to reach the home position as quickly as possible, so that we shall have a time-optimal problem. It is clear that an increase in the size of the applied force will speed up the motion, so that if the system is controllable the cost can be made arbitrarily small by taking the force arbitrarily large. Hence, we only obtain a sensible problem if the control is bounded.

With a suitable choice of units and of any parameters that may occur, we can write the state equations in the form

$$\dot{x}_1 = x_2, \quad \dot{x}_2 = u_1. \tag{1.13}$$

The first equation states that the velocity is the rate of change of position, and the second that the rate of change of velocity is proportional to the force. The control variable is to be integrable and bounded, so that

$$-1 \leq u_1(t) \leq +1 \quad \text{for all } t. \tag{1.14}$$

The initial time can be taken to be zero, and the initial and final states of the system are given by

$$x_1(0) = X, \quad x_2(0) = 0, \tag{1.15}$$

$$x_1(t_1) = 0, \quad x_2(t_1) = 0, \tag{1.16}$$

so that the problem is a fixed-target, free-time one. The cost is given by

$$J = \int_0^{t_1} 1 \mathrm{d}t = t_1, \tag{1.17}$$

and the problem is to minimize t_1.

It is obvious that, if X is positive, we should begin by applying a negative value of the control in order to move the object towards the origin. Also, we shall at some stage have to apply a positive control to slow the object down. It is also plausible that, in order to speed up the process, we should always use the largest control values available, that is, we should set the control to take values -1 and $+1$ only. Beginning with $u_1 = -1$, the initial section of the trajectory can be found by solving the state equations (1.13) and the inital conditions (1.15). The solution is

$$x_2 = -t, \quad x_1 = X - \tfrac{1}{2}t^2. \tag{1.18}$$

At some time t_2 we switch the control to the value $+1$. The state of the system then is given by (1.18) with $t = t_2$ and the solution of (1.13) for $t > t_2$ is given by

$$x_2 = t - 2t_2, \quad x_1 = \tfrac{1}{2}t^2 - 2tt_2 + X + t_2^2. \tag{1.19}$$

It is therefore *possible* to reach the origin at time t_1, provided we can find values of t_1 and t_2 such that

$$t_1 - 2t_2 = 0, \tag{1.20}$$

$$\tfrac{1}{2}t_1^2 - 2t_1t_2 + X + t_2^2 = 0. \tag{1.21}$$

These equations have the solution

$$t_1 = 2X^{1/2}, \quad t_2 = \tfrac{1}{2}t_1. \tag{1.22}$$

The controls we have used to reach the target have been bang–bang and we have made only a single switch. It is possible to reach the target in many other ways, with multiple switchings and with smoothly varying controls. We shall prove later that the solution we have obtained is, in fact, the optimal solution.

Finally, it is clear that if the initial position of the object is on the opposite side of the target, that is, if X is negative, a reversal of the strategy will enable us to reach the origin: start with $u_1 = +1$ and switch to $u_1 = -1$.

The positioning problem we have considered is mathematically identical to a classic problem in optimal control theory known as the rocket car. Here a tracked vehicle supports rocket engines so that a thrust in either direction along the track can be applied to the vehicle. There are obvious extensions to the problem of manoeuvring a spacecraft, although then the amount of fuel used up in the positioning of the craft is of greater importance than the time required. It is possible to construct a simple model that exhibits the same features. Take a grooved piece of wood to act as a runner and pivot it about a horizontal axis through its centre at right angles to the groove. Fix stops to permit the runner to rotate through a small angle above and below the horizontal. Place a marble on the groove, and the task is to adjust the slope of the wood so that the marble comes to rest at the pivot and to reach this position in the shortest possible time. The strategy that we have found above for this motion is to tilt the runner as much as possible, and at a precise moment swing it to its opposite extreme position. When the marble reaches the centre, which it should do with zero speed, turn the runner to the horizontal position, and it will remain at rest there.

1.5. Diabetes mellitus

In applying optimal control theory, the primary task, and often the most difficult, is to formulate an adequate model of the system under consideration. The model must include only the processes deemed to be most important in the working of the system. To attempt a *complete* model of all the components of the system would usually give a formulation too complicated for analysis. As an example of what must be done, consider the medical condition known as *diabetes mellitus.*

The amount of glucose in the blood is controlled by the presence of the hormone insulin. This is produced in the pancreas and its presence enables the excess glucose to be deposited in the liver and other tissues. Its release into the bloodstream is naturally controlled by the amount of glucose in the blood, so that in a healthy patient the ingestion of glucose in food or drink triggers an increase in the insulin level which enables the glucose to be removed. The insulin degenerates under natural metabolic processes so that a continuous source of supply is needed. If there is a deficiency in the supply of insulin from the pancreas, or an inadequate response of insulin production to meet high glucose levels, serious deleterious effects on the health of the patient will develop. If the natural controlling mechanism fails, it can be replaced by an artificial one by the administration of insulin by injection. The whole process is more complicated than this short description suggests, but the simplified version contains the key elements and is describable in terms of two variables. It is illustrated schematically in Fig. 1.1. In this figure, the solid lines denote the passage of material (glucose or hormone) and the broken lines the reciprocal influence of each component on the other.

Denote the amounts of glucose and hormone present at time t by the functions $G(t)$ and $H(t)$, respectively. Then the state equations have the form

$$\dot{G}(t) = F_1(G,H) + P(t), \tag{1.23}$$

$$\dot{H}(t) = F_2(G,H) + U(t), \tag{1.24}$$

where $P(t)$ denotes the rate of increase of glucose in the blood produced by food or drink, and $U(t)$ the rate of increase of insulin produced by its injection. These equations exhibit two mechanisms for the control of the glucose level: diet, which changes the value of P, and medication, which alters U. Suppose the equilibrium levels in

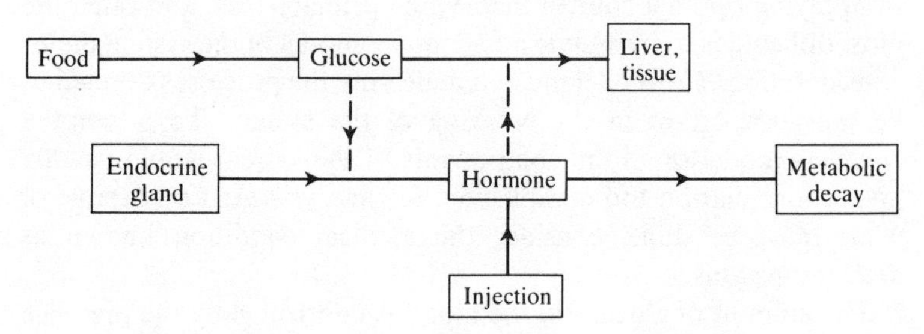

Fig. 1.1. Schematic diagram of the glucose–hormone system.

a healthy state are G_0 and H_0 and we consider small departures g and h from these values. Thus we define

$$g(t) = G(t) - G_0, \quad h(t) = H(t) - H_0, \tag{1.25}$$

and write the state equations in linearized form, neglecting squares of the small quantities. We obtain state equations of the form

$$\dot{g} = -c_1 g - c_2 h + p, \quad \dot{h} = -c_3 h + c_4 g + u, \tag{1.26}$$

where c_i, $i = 1, 2, 3, 4$, can be found from the partial derivatives of F and G, evaluated at the equilibrium level, and the signs have been chosen to exhibit the appropriate effect of each term, with c_i positive. In the pathological condition, the coefficient c_4 is greatly reduced or is zero. Suppose we start at $t = 0$ immediately after a meal. The initial levels of g and h are g_0 and h_0, say, and we wish to reduce them both to zero, that is, to reach the equilibrium levels, at time T without any further intake of food. Thus we can set $p(t) = 0$, and the initial and final conditions are

$$g(0) = g_0, \quad h(0) = h_0, \tag{1.27}$$

$$g(T) = 0, \quad h(T) = 0. \tag{1.28}$$

Other sets of conditions are possible. We might want to insist that the hormone level returns to its initial value; for example, we might have $h_0 = 0$. Or we might be content with reducing the excess glucose level to zero without imposing any condition on the final hormone level, dropping the second condition in (1.28).

To complete the specification of the problem, we need to define a suitable cost function. We wish to reduce the excess amount of glucose in the blood, since this has harmful effects. Also, we wish to use the minimum amount of drug consistent with the desired aim of the treatment, both because of financial constraints and because there may be harmful side-effects if the dosage is high. A cost function that accounts for both these penalties is given by

$$J = \int_0^T \tfrac{1}{2}(g^2 + k^2 u^2)\,dt, \tag{1.29}$$

where k^2 is a weighting factor between the two components of the cost.

We have now specified an optimal control problem for the control of diabetes, albeit in an over-simplified form. We are not able, at this stage, to proceed further with the solution. It is an example of an important class of optimal control problems, that of linear state equations with quadratic cost functions. Such problems will be considered in detail in Chapter 8. The aim of the discussion of this example here has been to show some of the steps that are necessary in moving from a particular system and its possible behaviour to a mathematical model and the statement of the corresponding optimal control problem in a standard form.

Before discussing methods for solving optimal control problems, it is convenient to collect together some mathematical methods and concepts that will be needed later, and these will be found in the next chapter.

Exercises 1

(Solutions on p. 217)

1.1 Consider the growth equation

$$\dot{x} = tu,$$

with $x(0) = 0$, $x(1) = 1$, and with the cost function

$$J = \int_0^1 u^2 \, \mathrm{d}t.$$

Show that $u^* = 3t$ is a successful control, with $x^* = t^3$ and $J^* = 3$ the corresponding trajectory and cost. If $u = u^* + v$ is another successful control, show that

$$\int_0^1 vt \, \mathrm{d}t = 0,$$

and, by finding the cost for this control, show that u^* is the optimal control.

1.2 The positioning problem is defined in §1.4. Suppose that the initial conditions are changed to $x_1(0) = 0$, $x_2(0) = X$, with $X > 0$. Show that a bang–bang control with a single switch can be employed to steer the system to the origin. Find the time taken, and also when and where the switch must be made. (As in §1.4, this is, in fact, the optimal solution, as will be proved in §4.4(2) and §6.3(3).)

1.3 Formulate an optimal control problem by thinking of a system that develops with time. Identify the relevant variables, a suitable form for the state equations, possible control variables for the system, and an appropriate cost function. Some examples of the construction of models for a variety of systems have been given in this chapter, and more will be found in later chapters. It is instructive to attempt such a formulation for yourself, even though you may not be able to solve the resulting equations at present.

2 Systems of differential equations, matrices, and sets

> Grow old along with me! The best is yet to be.
>
> Robert Browning

This chapter contains some material that will be used extensively in what follows. For convenience, it is collected together in this preliminary chapter, before we go on to consider optimal control problems. Some of the topics will be introduced briefly and some will be quoted without proof. If needed, further information can be found in the books listed in the Bibliography.

2.1. Ordinary differential equations

Because we shall be dealing with deterministic and evolutionary systems, the governing equations will be in the form of ordinary differential equations, with time as the independent variable. If the system is n-dimensional, the state of the system will be defined by an n-dimensional vector $\mathbf{x}$, and the state equations will have the form

$$\dot{\mathbf{x}} = \mathbf{f}(t, \mathbf{x}), \tag{2.1}$$

where $\mathbf{f}$ is also n-dimensional. The components of $\mathbf{f}$ will depend on some control variable $\mathbf{u}$ when we use these results, but we can suppress this information here. Note that the equations are of the first order. If the system is posed in terms of differential equations of second or higher order, we can always replace it by a system containing first-order equations only, but with more variables. For example, the second-order equation

$$\frac{\mathrm{d}^2x}{\mathrm{d}t^2} = -k^2x + u \tag{2.2}$$

can be replaced by a first-order system with two variables $x_1 = kx$ and $x_2 = \mathrm{d}x/\mathrm{d}t$. The state equations can be written as

$$\begin{aligned} \dot{x}_1 &= kx_2, \\ \dot{x}_2 &= -kx_1 + u. \end{aligned} \tag{2.3}$$

The right-hand sides of these two equations give the two components of $\mathbf{f}$.

The initial-value problem for the system (2.1) is to determine $\mathbf{x}(t)$, given that $\mathbf{x}(t)$ satisfies (2.1) and the initial condition

$$\mathbf{x}(t_0) = \mathbf{x}^0. \tag{2.4}$$

Picard's theorem states that the solution exists in some neighbourhood of t_0 if $\mathbf{f}(t, \mathbf{x})$ is continuous in its $n + 1$ variables, and that the solution is unique if $\mathbf{f}$ satisfies a Lipschitz condition. This condition requires that

$$\| \mathbf{f}(t, \mathbf{y}) - \mathbf{f}(t, \mathbf{z}) \| \leq K \| \mathbf{y} - \mathbf{z} \| \tag{2.5}$$

for some positive constant K and for all t near t_0 and all $\mathbf{y}$ and $\mathbf{z}$ near $\mathbf{x}^0$. The norm used in this condition is the Euclidean norm, that is, the length of the vector, so that

$$\| \mathbf{x} \|^2 = \mathbf{x} \,.\, \mathbf{x} = \mathbf{x}^{\mathrm{T}}\mathbf{x}, \tag{2.6}$$

where the dot denotes the scalar product and the superscript T the transpose.

We can be less certain about the solution of the boundary-value problem; this occurs when the state vector is not completely prescribed at the initial instant, and instead we require it to satisfy certain conditions at a later time. Existence and uniqueness are not assured in this case, even for some very simple systems. For example, consider the system (2.3) with $u = 0$ and $t_0 = 0$, and suppose also that $x_1(0) = 0$. The solution at time t is then easily found to have the form

$$x_1 = A \sin kt, \quad x_2 = A \cos kt, \tag{2.7}$$

where A is some arbitrary constant which is equal to the initial value of x_2. If this is given, as it is in an initial-value problem, the solution exists and is unique. However, if instead we require x_1 to have the value 1 when $kt = \pi$, it is clear that no such solution exists. Further, if instead we require that x_1 should be zero at this time, we see from

(2.7) that this condition is met whatever value A takes, and so we have an infinite number of solutions. This distinction between initial- and boundary-value problems for systems of ordinary differential equations will be seen later to be important.

2.2. Linear differential equations

There are no general methods by which the solution of a nonlinear system of differential equations can be found. For linear systems, however, this is possible. Consider first the one-dimensional linear equation, which has the general form

$$\dot{x} = A(t)x + b(t), \quad x(t_0) = x^0. \tag{2.8}$$

If $b \equiv 0$, the solution can be found by separation of variables, and it is easily seen that

$$x(t) = X(t, t_0)x^0, \tag{2.9}$$

where

$$X(t, t_0) = \exp\left(\int_{t_0}^{t} A(\tau)\,\mathrm{d}\tau\right). \tag{2.10}$$

The reciprocal of $X(t, t_0)$ is $X(t_0, t)$ and can be used as an integrating factor for the full equation. The solution, as can easily be verified, is given by

$$x(t) = X(t, t_0)\left(x^0 + \int_{t_0}^{t} X(t_0, \tau)b(\tau)\,\mathrm{d}\tau\right). \tag{2.11}$$

This is a general statement of a familiar elementary method for solving differential equations. The question we would like to answer is whether there exists a similar formalism for linear n-dimensional equations.

The general n-dimensional linear equation can be written in the form

$$\dot{\mathbf{x}} = \mathbf{A}(t)\mathbf{x} + \mathbf{b}(t), \quad \mathbf{x}(t_0) = \mathbf{x}^0. \tag{2.12}$$

The homogeneous equation with $\mathbf{b} = \mathbf{0}$ has a solution of the form

$$\mathbf{x}(t) = \mathbf{X}(t, t_0)\mathbf{x}^0, \tag{2.13}$$

where $\mathbf{X}$ is an $n \times n$ matrix, known as the *fundamental matrix*. To construct $\mathbf{X}$ we can proceed as follows. Determine the n solutions of the homogeneous equations with initial values equal to one of the unit vectors $\mathbf{e}^{(i)}$, $i = 1, 2, \ldots, n$, which are the columns of the unit matrix. These solutions exist and are unique because of Picard's theorem. If they are written as $\mathbf{x}^{(i)}(t, t_0)$, we can put them together as the columns of a matrix, which is the fundamental matrix $\mathbf{X}$. For it is clear that $\mathbf{X}(t_0, t_0) = \mathbf{I}$, the unit matrix, and since each column satisfies the homogeneous form of the differential equation (2.1) so does the whole matrix. Thus

$$\dot{\mathbf{X}} = \mathbf{AX} \tag{2.14}$$

and (2.13) is easily verified to be the solution of (2.12) with $\mathbf{b} = \mathbf{0}$.

Let us find the fundamental matrix for the harmonic oscillator. The equation to be solved has the form

$$\ddot{x} + k^2 x = 0, \tag{2.15}$$

which as we have seen can be written as a pair of first-order equations by defining $x_1 = x$ and $x_2 = \dot{x}/k$. The equation (2.15) can then be written in matrix form as

$$\begin{bmatrix} \dot{x}_1 \\ \dot{x}_2 \end{bmatrix} = \begin{bmatrix} 0 & k \\ -k & 0 \end{bmatrix} \begin{bmatrix} x_1 \\ x_2 \end{bmatrix}. \tag{2.16}$$

The solutions of these equations that satisfy the conditions $x_1 = 1$, $x_2 = 0$, and $x_1 = 0$, $x_2 = 1$, at $t = t_0$ are easily found and, with these solutions forming the columns of the fundamental matrix, we see that in this case

$$\mathbf{X}(t, t_0) = \begin{bmatrix} \cos k(t - t_0) & \sin k(t - t_0) \\ -\sin k(t - t_0) & \cos k(t - t_0) \end{bmatrix}. \tag{2.17}$$

We can now prove some general results concerning the fundamental matrix. First we have the transition property,

$$\mathbf{X}(t_2, t_0) = \mathbf{X}(t_2, t_1)\mathbf{X}(t_1, t_0). \tag{2.18}$$

Let $\mathbf{x}^0$ be an arbitrarily chosen initial value for $\mathbf{x}$ at $t = t_0$. Then repeated application of the general result (2.13) shows that

$$\mathbf{x}(t_1) = \mathbf{X}(t_1, t_0)\mathbf{x}^0, \quad \mathbf{x}(t_2) = \mathbf{X}(t_2, t_1)\mathbf{x}(t_1), \quad \mathbf{x}(t_2) = \mathbf{X}(t_2, t_0)\mathbf{x}^0. \tag{2.19}$$

If we combine the first two expressions and compare the result with the third, the transition property (2.18) follows since $\mathbf{x}^0$ is arbitrary. For the second result, set $t_2 = t_0$ in (2.18). Then, since

$$\mathbf{X}(t_0, t_0) = \mathbf{I}, \tag{2.20}$$

$$\mathbf{X}(t_0, t_1)\mathbf{X}(t_1, t_0) = \mathbf{I}. \tag{2.21}$$

Two important consequences of this result are that the fundamental matrix is never singular and its inverse is given by reversing the order of the two arguments, that is,

$$\mathbf{X}(t, t_0)^{-1} = \mathbf{X}(t_0, t). \tag{2.22}$$

We have already seen that the derivative of the fundamental matrix $\mathbf{X}(t, t_0)$ with respect to t satisfies the same differential equation as $\mathbf{x}$, namely

$$\dot{\mathbf{X}} = \mathbf{AX}. \tag{2.23}$$

To find the derivative of the inverse of $\mathbf{X}$ we start from (2.21) with t_1 replaced by t and differentiate with respect to t. We then find that

$$\dot{\mathbf{X}}(t_0, t)\mathbf{X}(t, t_0) + \mathbf{X}(t_0, t)\dot{\mathbf{X}}(t, t_0) = \mathbf{0}. \tag{2.24}$$

This equation can be simplified by making use of (2.23) and post-multiplying by $\mathbf{X}(t, t_0)^{-1}$, which gives

$$\dot{\mathbf{X}}(t_0, t) = -\mathbf{X}(t_0, t)\mathbf{A}(t), \tag{2.25}$$

or

$$\frac{\mathrm{d}\mathbf{X}^{-1}}{\mathrm{d}t} = -\mathbf{X}^{-1}\mathbf{A}. \tag{2.26}$$

Note that the inverse of the time derivative of $\mathbf{X}$, from (2.23), is equal to $\mathbf{X}^{-1}\mathbf{A}^{-1}$.

We can make use of this last result to construct the solution of the inhomogeneous equation (2.12). Define a new variable $\mathbf{y}(t)$ by

$$\mathbf{y}(t) = \mathbf{X}(t_0, t)\mathbf{x}(t) = \mathbf{X}^{-1}\mathbf{x}. \tag{2.27}$$

Then

$$\begin{aligned}\dot{\mathbf{y}} &= \dot{\mathbf{X}}(t_0, t)\mathbf{x}(t) + \mathbf{X}(t_0, t)\dot{\mathbf{x}}(t)\\ &= -\mathbf{X}^{-1}\mathbf{Ax} + \mathbf{X}^{-1}(\mathbf{Ax} + \mathbf{b})\\ &= \mathbf{X}^{-1}\mathbf{b}.\end{aligned} \tag{2.28}$$

From the definition of $\mathbf{y}$ we see that $\mathbf{y}(t_0) = \mathbf{x}^0$ and the solution of (2.28) follows by direct integration. We find that

$$\mathbf{y}(t) = \int_{t_0}^{t} \mathbf{X}(t_0, \tau)\mathbf{b}(\tau)\,\mathrm{d}\tau + \mathbf{x}^0, \tag{2.29}$$

and since from the definition of $\mathbf{y}$ it follows that $\mathbf{x} = \mathbf{X}\mathbf{y}$, we finally obtain the general solution of the inhomogeneous linear system (2.12) in the form

$$\mathbf{x}(t) = \mathbf{X}(t, t_0)\left(\mathbf{x}^0 + \int_{t_0}^{t} \mathbf{X}(t_0, \tau)\mathbf{b}(\tau)\,\mathrm{d}\tau\right). \tag{2.30}$$

Notice that this solution has exactly the same form as that for the one-dimensional case, given by (2.11). The difference is that there it was possible to find X as an explicit integral, but the determination of the fundamental matrix when $n \neq 1$ is not so straightforward. There is, however, a very important special case in which the fundamental matrix can be easily determined and this we proceed to examine.

2.3. Autonomous systems

By an *autonomous system* we mean one in which time does not appear explicitly in the differential equation governing the evolution of the state of the system. For a linear system, we can relax this requirement somewhat and all we need do is restrict attention to equations of the form (2.12) in which $\mathbf{A}$ is a constant matrix, but $\mathbf{b}$ can remain dependent on t. Because t is not present explicitly in the homogeneous system

$$\dot{\mathbf{x}} = \mathbf{A}\mathbf{x}, \tag{2.31}$$

the solution and the fundamental matrix will depend on the time interval $t - t_0$ only. (See, for example, the matrix (2.17) for the autonomous system (2.16).) Hence, for autonomous systems $\mathbf{X}(t, t_0) = \mathbf{X}(t - t_0, 0)$ and it follows that, with no loss of generality, we can set $t_0 = 0$. It also follows that $\mathbf{X}(t_0, t) = \mathbf{X}(t_0 - t, 0)$, and so, when $t_0 = 0, \mathbf{X}(0, t) = \mathbf{X}(-t, 0)$.

When $\mathbf{A}$ is a constant matrix, the fundamental matrix has the simple explicit form

$$\mathbf{X}(t, 0) = \exp(\mathbf{A}t). \tag{2.32}$$

The exponential function of a matrix is defined by an infinite series of identical form to that for $\exp(t)$. To establish (2.32), we have to show that the solution of the differential equation $\dot{\mathbf{X}} = \mathbf{AX}$, with $\mathbf{X}(0, 0) = \mathbf{I}$, has the form

$$\mathbf{X} = \mathbf{I} + \sum_{k=1}^{\infty} \frac{\mathbf{A}^k t^k}{k!}. \tag{2.33}$$

By using a suitable matrix norm, we can justify the differentiation of this series term by term, and then it at once follows that the differential equation for $\mathbf{X}$ is satisfied. It is also obvious that the initial condition is satisfied and hence we have established the form (2.32) for the fundamental matrix. It follows at once that the solution of the inhomogeneous problem, given by (2.30) in the general case, now has the form

$$\mathbf{x}(t) = \exp(\mathbf{A}t)\left(\mathbf{x}^0 + \int_0^t \exp(-\mathbf{A}\tau)\mathbf{b}(\tau)\,\mathrm{d}\tau\right). \tag{2.34}$$

As an example of how the fundamental matrix can be found, consider the case when

$$\mathbf{A} = \begin{bmatrix} 0 & 1 \\ 1 & 0 \end{bmatrix}. \tag{2.35}$$

Then $\mathbf{A}^2 = \mathbf{I}$ and it follows that $\mathbf{A}^{2k} = \mathbf{I}$ and $\mathbf{A}^{2k+1} = \mathbf{A}$ for all integers k. Hence, from (2.33),

$$\begin{aligned} \exp(\mathbf{A}t) &= \mathbf{I}\left(1 + \sum_1^{\infty} \frac{t^{2k}}{(2k)!}\right) + \mathbf{A}\sum_1^{\infty} \frac{t^{2k-1}}{(2k-1)!} \\ &= \begin{bmatrix} \cosh t & \sinh t \\ \sinh t & \cosh t \end{bmatrix}. \end{aligned} \tag{2.36}$$

A similar method can be used for other matrices $\mathbf{A}$.

2.4. The adjoint system

Returning to the general case when $\mathbf{A}$ is a function of t, we can define the *adjoint system* as the system with matrix $-\mathbf{A}^{\mathrm{T}}$, where the superscript T denotes the transpose of the matrix. If $\mathbf{Y}$ is the fundamental matrix for the adjoint system, the result (2.14) applied to both systems gives

$$\dot{\mathbf{X}} = \mathbf{AX}, \quad \dot{\mathbf{Y}} = -\mathbf{A}^{\mathrm{T}}\mathbf{Y}. \tag{2.37}$$

If we transpose the second of these equations we find that

$$\dot{\mathbf{Y}}^{\mathrm{T}} = -\mathbf{Y}^{\mathrm{T}}\mathbf{A}, \tag{2.38}$$

and it follows that

$$\frac{\mathrm{d}}{\mathrm{d}t}(\mathbf{Y}^{\mathrm{T}}\mathbf{X}) = \mathbf{Y}^{\mathrm{T}}\dot{\mathbf{X}} + \dot{\mathbf{Y}}^{\mathrm{T}}\mathbf{X} = \mathbf{0}. \tag{2.39}$$

Hence $\mathbf{Y}^{\mathrm{T}}\mathbf{X}$ is a constant matrix and since, by definition, $\mathbf{X}$ and $\mathbf{Y}$ are both equal to $\mathbf{I}$ when $t = t_0$,

$$\mathbf{Y}^{\mathrm{T}}\mathbf{X} = \mathbf{I}, \tag{2.40}$$

or equivalently,

$$\mathbf{Y}(t, t_0) = \mathbf{X}^{\mathrm{T}}(t_0, t). \tag{2.41}$$

The consequence of this result is that, if the values of $\mathbf{x}$ and $\mathbf{y}$ at t_0 are $\mathbf{x}^0$ and $\mathbf{y}^0$ respectively,

$$\mathbf{x}(t) = \mathbf{X}(t, t_0)\mathbf{x}^0, \quad \mathbf{y}(t) = \mathbf{Y}(t, t_0)\mathbf{y}^0, \tag{2.42}$$

and

$$\mathbf{y}^{\mathrm{T}}\mathbf{x} = \mathbf{y}^{0\mathrm{T}}\mathbf{Y}^{\mathrm{T}}\mathbf{X}\mathbf{x}^0 = \mathbf{y}^{0\mathrm{T}}\mathbf{x}^0. \tag{2.43}$$

Thus, the scalar product of the solutions of the system and its adjoint does not change with time.

A system for which $\mathbf{A} = -\mathbf{A}^{\mathrm{T}}$ is called *self-adjoint*. In this case, $\mathbf{Y} = \mathbf{X}$ and

$$\mathbf{X}^{\mathrm{T}}\mathbf{X} = \mathbf{I}, \tag{2.44}$$

so the fundamental matrix is orthogonal. The invariance of the scalar product of the solutions of the system and its adjoint now becomes

$$\mathbf{x}^{\mathrm{T}}\mathbf{x} = \mathbf{x}^{0\mathrm{T}}\mathbf{x}^0, \quad \text{or} \quad \|\mathbf{x}\| = \|\mathbf{x}^0\|, \tag{2.45}$$

and the length of the solution vector is constant. The condition that a system be self-adjoint is that the matrix $\mathbf{A}$ be skew-symmetric; an example is the matrix for the harmonic oscillator given in (2.16) and the corresponding fundamental matrix given by (2.17) is clearly orthogonal.

2.5. Vectors and matrices

We have already made use of some properties of matrices, but we list here some elementary definitions and results that will be important in what follows.

As well as the Euclidean norm of a vector already defined, we also use the vector and matrix norms defined by

$$|\mathbf{x}| = \sum_i |x_i|, \quad |\mathbf{A}| = \sum_{i,j} |A_{ij}|, \tag{2.46}$$

where x_i and A_{ij} are the elements of the vector $\mathbf{x}$ and the matrix $\mathbf{A}$, respectively. The *rank* of the matrix is the maximum number of linearly independent rows (or columns) in the matrix. For a square matrix of dimension n, the *eigenvalues* λ are the roots of the equation

$$\det(\lambda\mathbf{I} - \mathbf{A}) = 0, \tag{2.47}$$

which is known as the *characteristic equation* of $\mathbf{A}$. If zero is an eigenvalue of the matrix, the rank of the matrix is less than n and the matrix is said to be *singular*. If the rank is equal to k, there are k linearly independent columns in $\mathbf{A}$ and it is possible to find $n - k$ linearly independent vectors, each orthogonal to all the columns of $\mathbf{A}$.

The Cayley–Hamilton theorem states that a matrix satisfies its own characteristic equation. This means that if λ is replaced by $\mathbf{A}$ in the polynomial that results from expanding the left-hand side of (2.47) we obtain the zero matrix $\mathbf{0}$. It follows that $\mathbf{A}^n$ is linearly dependent on $\mathbf{I}$ and $\mathbf{A}^k$, for $k = 1, 2, \ldots, n - 1$. Moreover, it follows easily by induction that $\mathbf{A}^m$ is also dependent on the same n matrices for all values of m greater than n. A consequence of this theorem is that the fundamental matrix $\exp(\mathbf{A}t)$ defined by the infinite series (2.33) can be expressed as a polynomial in the matrices $\mathbf{I}, \mathbf{A}, \mathbf{A}^2, \ldots, \mathbf{A}^{n-1}$, which helps in its evaluation in some cases.

If a matrix $\mathbf{A}$ has n distinct eigenvalues $\lambda_1, \lambda_2, \ldots, \lambda_n$, there exists a non-singular matrix $\mathbf{P}$ such that

$$\mathbf{P}^{-1}\mathbf{A}\mathbf{P} = \mathbf{\Lambda}, \tag{2.48}$$

where $\mathbf{\Lambda}$ is the diagonal matrix with elements equal to the eigenvalues. When the eigenvalues are not all distinct, such a reduction may still be possible, but otherwise the best that can be achieved is reduction to a form in which there are some non-zero entries in the diagonal above the main diagonal.

The eigenvalues of a real symmetric matrix are all real. Since the matrix $\lambda\mathbf{I} - \mathbf{A}$ is singular, the equation

$$\mathbf{A}\mathbf{x} = \lambda\mathbf{x} \tag{2.49}$$

has a non-trivial solution which is called the *eigenvector*. The eigenvectors corresponding to distinct eigenvalues are orthogonal, that is, if

$$\mathbf{Ax} = \lambda\mathbf{x}, \quad \mathbf{Ay} = \mu\mathbf{y}, \tag{2.50}$$

where $\lambda \neq \mu$, then

$$\mathbf{y}^{\mathrm{T}}\mathbf{x} = 0. \tag{2.51}$$

If $\mathbf{A}$ is a real symmetric matrix with elements a_{ij}, the scalar quantity $f = \mathbf{y}^{\mathrm{T}}\mathbf{Ax}$, where $\mathbf{x}$ and $\mathbf{y}$ are n-dimensional vectors is a *bilinear form*, given by the double sum

$$f = \sum_{i,j} a_{ij} y_i x_j. \tag{2.52}$$

If we differentiate this expression, we find that

$$\frac{\partial f}{\partial y_i} = \sum_j a_{ij} x_j = (\mathbf{Ax})_i, \tag{2.53}$$

or

$$\frac{\partial f}{\partial \mathbf{y}} = \mathbf{Ax}. \tag{2.54}$$

If we transpose the scalar quantity f, we see that $f = \mathbf{x}^{\mathrm{T}}\mathbf{A}^{\mathrm{T}}\mathbf{y} = \mathbf{x}^{\mathrm{T}}\mathbf{Ay}$, since $\mathbf{A}$ is symmetric, and so

$$\frac{\partial f}{\partial \mathbf{x}} = \mathbf{Ay}. \tag{2.55}$$

If $\mathbf{y} = \mathbf{x}$, f is a *quadratic form*, and

$$\frac{\partial f}{\partial \mathbf{x}} = 2\mathbf{Ax}. \tag{2.56}$$

A quadratic form f is *positive definite* if $f > 0$ for all $\mathbf{x}$ except $\mathbf{x} = \mathbf{0}$. The condition for a quadratic form to be positive definite is that each of the eigenvalues of the matrix $\mathbf{A}$ of the form be positive. (We have already noted that all the eigenvalues are real.) The form is *positive semi-definite* if $f \geq 0$ for all $\mathbf{x}$, and this will be so if the eigenvalues are all non-negative. The form $x_1^2 + x_1x_2 + x_2^2$ is positive definite, but $x_1^2 + 2x_1x_2 + x_2^2$ is only semi-definite.

2.6. Convex sets

In linear optimal control problems, the solution of state equations of the form (2.12) is often required, but the forcing term $\mathbf{b}(t)$ is not predetermined because it depends on the particular control being used. With a given initial state for the system, we can therefore arrive at a collection of possible states that can be reached in a given time. These possible states form a set $\mathscr{S} \subset \mathscr{R}^n$, the set of n-dimensional vectors with real elements. Some of the properties of these sets, which will be needed in what follows, are briefly listed here.

The set is *closed* if the limits of all convergent sequences of points in $\mathscr{S}$ belong to $\mathscr{S}$. The set $\mathscr{S}$ is *convex* if all points lying on the line joining two points $\mathbf{x}$ and $\mathbf{y}$ belonging to $\mathscr{S}$ also belong to $\mathscr{S}$; if $\mathbf{x}, \mathbf{y} \in \mathscr{S}$, then $c\mathbf{x} + (1 - c)\mathbf{y} \in \mathscr{S}$ for $0 < c < 1$. The *interior* of the set, $\text{Int}(\mathscr{S})$, are those points $\mathbf{x} \in \mathscr{S}$ such that it is possible to find a ball $\mathscr{B}(\mathbf{x}, r)$, centre $\mathbf{x}$ and radius r, which lies entirely within $\mathscr{S}$. If $\mathscr{S} = \text{Int}(\mathscr{S})$, the set is *open*. If all balls with centre $\mathbf{x}$ contain points *not* belonging to $\mathscr{S}$, then $\mathbf{x}$ belongs to the *boundary* of the set, $\partial\mathscr{S}$. Formally, these definitions can be stated as

$$\text{Int}(\mathscr{S}) = \{\mathbf{x} \in \mathscr{S} \mid \mathscr{B}(\mathbf{x}, r) \subset \mathscr{S}\} \quad \text{for some } r > 0, \tag{2.57}$$

$$\partial\mathscr{S} = \{\mathbf{x} \notin \text{Int}(\mathscr{S}) \quad \text{and} \quad \mathbf{x} \notin \text{Int}(\mathscr{S}^c)\}, \tag{2.58}$$

where $\mathscr{S}^c$ is the complement of $\mathscr{S}$. The set $\mathscr{S}$ is said to be *strictly convex* if the line joining two points in the set lies in the interior of the set, that is, $c\mathbf{x} + (1 - c)\mathbf{y} \in \text{Int}(\mathscr{S})$ for $0 < c < 1$. The diagrams in Fig. 2.1 show two-dimensional examples of sets that are not convex, convex, and strictly convex.

An *extreme* point $\mathbf{q} \in \mathscr{S}$ is one that does *not* lie on the line joining two different points of $\mathscr{S}$. Thus $\mathbf{x}$ is extreme if it is not possible to find two other points of $\mathscr{S}$, $\mathbf{y}$ and $\mathbf{z}$ say, such that $\mathbf{x} = \frac{1}{2}(\mathbf{y} + \mathbf{z})$. It is clear that no interior point can be extreme, and that all points on the

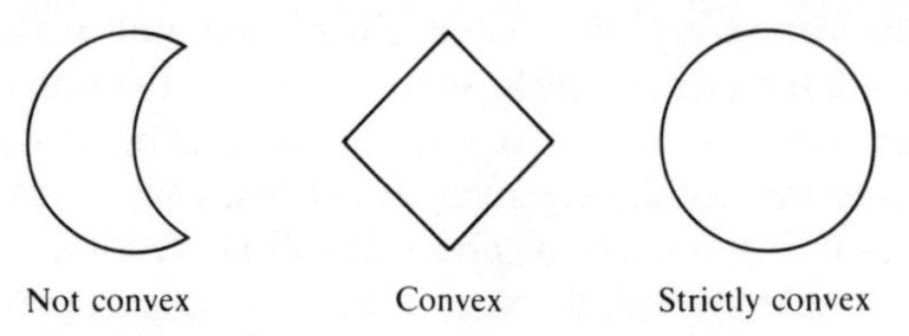

Fig. 2.1. Types of two-dimensional sets.

boundary of a strictly convex set are extreme. For a set that is convex, but not strictly so, corner points are extreme, but points on the flat portions of the boundary are not extreme. All points on the circular edge of a disc are extreme, and the only extreme points of a triangle are the vertices.

In solving optimal control problems, we shall frequently have occasion to determine the greatest value of a bounded real function f defined for all points $\mathbf{x}$ lying in some set $\mathscr{S} \subset \mathscr{R}^n$. This greatest value is called the *supremum* of f over $\mathscr{S}$. We write

$$f_{\text{sup}} = \sup_{\mathbf{x} \in \mathscr{S}} f(\mathbf{x}) \tag{2.59}$$

and f_{sup} has the properties that

$$f_{\text{sup}} \geq f(\mathbf{x}) \quad \text{for all } \mathbf{x} \in \mathscr{S}, \tag{2.60}$$

and, for any $\varepsilon > 0$,

$$f_{\text{sup}} - \varepsilon < f(\mathbf{x}) \quad \text{for some } \mathbf{x} \in \mathscr{S}. \tag{2.61}$$

Another name for the supremum is the least upper bound. Similarly, the smallest value of f for $\mathbf{x} \in \mathscr{S}$ is called the *infimum* of f over $\mathscr{S}$ and can be defined in a parallel way, with inequalities reversed.

As an example in $\mathscr{R}^2$, let $f = x_1^2 + x_2^2$ and choose $\mathscr{S}$ to be the elliptic disc $x_1^2 + 4x_2^2 \leq 4$. Then $f_{\text{sup}} = 4$ and $f_{\text{inf}} = 0$. If we take the supremum and infimum over $\partial\mathscr{S}$ instead of over $\mathscr{S}$, f_{sup} is unchanged, but $f_{\text{inf}} = 1$. It should be remembered that, when f is differentiable, the vanishing of the gradient of f may give *local* maxima or minima of f but these may or may not be equal to the supremum or infimum of f, which are the greatest and least values of f over the whole of the set of definition.

A *hyperplane* is a set $\mathscr{P}$ of dimension $n - 1$ defined by

$$\mathscr{P} = \{\mathbf{x} \in \mathscr{R}^n \,|\, \mathbf{a} \,.\, \mathbf{x} = p\} \tag{2.62}$$

for some scalar p and some n-dimensional vector $\mathbf{a}$, which is the *normal* to the hyperplane. If $\mathscr{S}$ is closed and bounded and $\mathbf{y} \notin \mathscr{S}$, there exists a hyperplane separating $\mathbf{y}$ from $\mathscr{S}$. If this hyperplane is defined by $\mathbf{a} \,.\, \mathbf{x} = p$, then $\mathbf{a} \,.\, \mathbf{x} < p$ for all $\mathbf{x} \in \mathscr{S}$ and $\mathbf{a} \,.\, \mathbf{y} > p$; see Fig. 2.2(a). If $\mathbf{y} \in \partial\mathscr{S}$, there exists a *supporting hyperplane* $\mathscr{P}$ through $\mathbf{y}$ such that $\mathscr{S}$ lies on one side of $\mathscr{P}$, that is $\mathbf{a} \,.\, \mathbf{y} = p$ and $\mathbf{a} \,.\, \mathbf{x} \leq p$ for $\mathbf{x} \in \mathscr{S}$; see Fig. 2.2(b). If $\mathscr{S}$ is strictly convex there is only one point common to both the plane and the set. However, if the boundary of

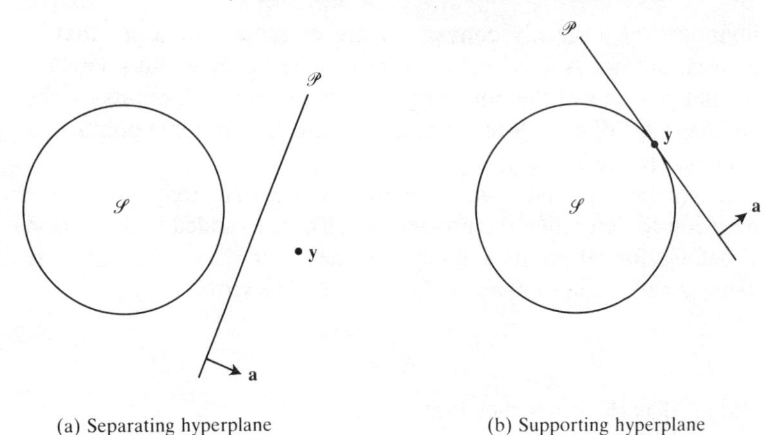

(a) Separating hyperplane (b) Supporting hyperplane

Fig. 2.2. Convex sets and hyperplanes in two dimensions.

$\mathscr{S}$ is not smooth the supporting hyperplane at a point on $\partial\mathscr{S}$ may not be unique. If $\mathscr{S}$ is not strictly convex, then a supporting hyperplane may be supporting for more than one boundary point. Some possible cases are shown in Fig. 2.3. Finally, the existence of the supporting hyperplane can be expressed formally as follows: for each $\mathbf{y} \in \partial\mathscr{S}$, there exists a normal vector $\mathbf{a}$ such that

$$\mathbf{a} \cdot \mathbf{y} = \sup_{\mathbf{x} \in \mathscr{S}} (\mathbf{a} \cdot \mathbf{x}). \tag{2.63}$$

On occasions we have to consider linear subspaces of $\mathscr{S}$ of dimension less than $n - 1$. These are formed by the intersection of

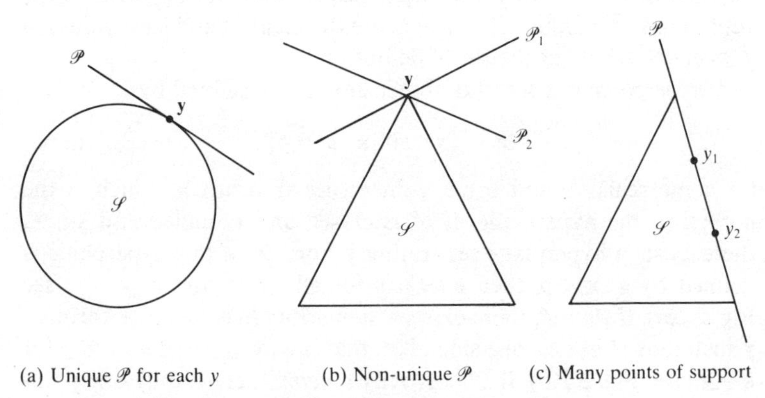

(a) Unique $\mathscr{P}$ for each y (b) Non-unique $\mathscr{P}$ (c) Many points of support

Fig. 2.3. Separating hyperplanes in two dimensions.

hyperplanes, and in $\mathscr{R}^3$ comprise lines and points. In higher dimensions, we shall refer to them generally as hyperplanes, giving the number of dimensions if this is not indicated by the context.

This concludes the discussion of some mathematical ideas that will be needed in what follows. We are now in a position to begin the study of optimal control. Either Part A or Part B can be studied first, and a general description of the contents of each part can be found on pages 29 and 81. The most logical order is to start with Part A, but if readers wish to learn quickly how to solve general optimal control problems, they should start with Part B.

Exercises 2

(*Solutions on p.* 217)

2.1 Justify the solution given in (2.11) for the one-dimensional equation (2.8).

2.2 Find the fundamental matrix for the two-dimensional system defined by

$$\dot{x}_1 = x_1 + tx_2, \quad \dot{x}_2 = x_2,$$

and determine the solution for which $x_1(0) = c_1, x_2(0) = c_2$.

2.3 Find the fundamental matrix for the autonomous systems in which the matrices $\mathbf{A}$ are given by

$$\text{(a) } \mathbf{A} = \begin{bmatrix} 1 & 1 \\ 0 & 0 \end{bmatrix}, \quad \text{(b) } \mathbf{A} = \begin{bmatrix} p & 1 \\ 0 & p \end{bmatrix},$$

$$\text{(c) } \mathbf{A} = \begin{bmatrix} 0 & 1 & 0 \\ 0 & 0 & 1 \\ 0 & 0 & 0 \end{bmatrix}, \quad \text{(d) } \mathbf{A} = \begin{bmatrix} 0 & 1 \\ 0 & -q \end{bmatrix},$$

(e) $\mathbf{A}$ is the diagonal matrix of order n, with elements $p_1, p_2, \ldots, p_n$.

2.4 Show that, in general, $\mathrm{d}(\mathbf{A}^2)/\mathrm{d}t \neq 2\mathbf{A}\dot{\mathbf{A}}$.

The fundamental matrix (2.10) for the non-autonomous one-dimensional system, and the matrix (2.32) for the n-dimensional autonomous system, suggest that the general case might be given by a result similar to (2.10), but with the scalar A replaced by the matrix $\mathbf{A}$. Show that this is not so, and explain why.

2.5 Let $\mathbf{X}$ be the fundamental matrix for the matrix $\mathbf{A}$ given in Exercise 2.3(a). Find the fundamental matrix $\mathbf{Y}$ for the adjoint system and verify that $\mathbf{Y} = (\mathbf{X}^{-1})^{\mathrm{T}}$.

2.6 Show that

$$\exp(\mathbf{A}t)\exp(-\mathbf{A}t) = \mathbf{I}.$$

Show also that

$$\exp(\mathbf{A}t)\exp(\mathbf{B}t) = \exp(\mathbf{A}t + \mathbf{B}t)$$

if and only if the matrices **A** and **B** commute.

2.7 Two matrices **A** and **B** are said to be *similar* if there exists a non-singular matrix **P** such that

$$\mathbf{A} = \mathbf{P}^{-1}\mathbf{B}\mathbf{P}.$$

If **A** and **B** are similar, show that $\exp(\mathbf{A}t)$ and $\exp(\mathbf{B}t)$ are also similar, with the same matrix **P**.

2.8 Show that the linear differential equation of order n, defined by

$$\frac{d^n y}{dt^n} + a_{n-1}\frac{d^{n-1}y}{dt^{n-1}} + \cdots + a_1\frac{dy}{dt} + a_0 y = f(t),$$

can be written as a linear system of n first-order equations. Find the matrix **A** of this system and show that its eigenvalues are the roots of the equation

$$p^n + a_{n-1}p^{n-1} + \cdots + a_1 p + a_0 = 0.$$

If the roots of this equation are distinct, find a matrix **P** such that

$$\mathbf{P}^{-1}\mathbf{A}\mathbf{P} = \mathbf{\Lambda},$$

where $\mathbf{\Lambda}$ is a diagonal matrix.

PART A
Time-optimal Control of Linear Systems

In Part A we discuss an important special case of optimal control problems, in which the state equations are linear and the cost is equal to the time. We shall find that it is possible to make some general statements about the optimal control and also to determine the solution in special cases explicitly. Before considering the optimal control problem itself, we first have to determine whether or not the system is controllable. This is done in Chapter 3. The main results about time-optimal control are in Chapter 4 and some examples are discussed in Chapter 5.

3 Controllability

Who can control his fate?

William Shakespeare

In this chapter we discuss the general question of the controllability of deterministic systems, mainly for linear systems, though some of the results are also true for nonlinear ones.

3.1. The controllable set

Suppose that we have a system defined by an n-dimensional state vector $\mathbf{x}$ and an m-dimensional control variable $\mathbf{u}$, with state equations of the form

$$\dot{\mathbf{x}} = \mathbf{f}(\mathbf{x}, \mathbf{u}), \tag{3.1}$$

where $\mathbf{f}$ is continuously differentiable in $\mathbf{x}$ and $\mathbf{u}$. We consider controllability to the origin, so that the target state is $\mathbf{x} = \mathbf{0}$, and we also suppose that $\mathbf{f}(\mathbf{0}, \mathbf{0}) = \mathbf{0}$. This ensures that, once we have reached the target, it is possible to remain there by switching off all the controls. The initial state is given by $\mathbf{x}(0) = \mathbf{x}^0$; since the time t does not occur explicitly in (3.1), we have an autonomous problem and can choose the initial time to be zero. The control variables are restricted to integrable functions of t, and we say that $\mathbf{u} \in \mathscr{U}_{\mathrm{u}}$ if the elements of $\mathbf{u}$ are without bounds, $\mathbf{u} \in \mathscr{U}_{\mathrm{b}}$ if $|u_i(t)| \leq 1, i = 1, 2, \ldots, m$, and $\mathbf{u} \in \mathscr{U}_{\mathrm{bb}}$ if $|u_i(t)| = 1$. It is clear that

$$\mathscr{U}_{\mathrm{bb}} \subset \mathscr{U}_{\mathrm{b}} \subset \mathscr{U}_{\mathrm{u}} \tag{3.2}$$

and we shall refer to these control sets as the bang–bang, bounded and unbounded cases. This defines the general nonlinear autonomous (NLA) problem. If we replace the state equations (3.1) by

$$\dot{\mathbf{x}} = \mathbf{A}\mathbf{x} + \mathbf{B}\mathbf{u}, \tag{3.3}$$

where $\mathbf{A}(n \times n)$ and $\mathbf{B}(n \times m)$ are constant matrices, we have the linear autonomous (LA) problem.

We define the *controllable set at time* t_1 as the set of initial states $\mathbf{x}^0$ that can be steered to the origin in time t_1 using an admissible control, that is, one belonging to the chosen control set. We can denote this set by $\mathscr{C}(t_1, \mathscr{U}, \mathbf{0})$ for controls $\mathbf{u} \in \mathscr{U}$, but usually we shall suppress the explicit dependence on the control set and target state and write it as $\mathscr{C}(t_1)$. The *controllable set* $\mathscr{C}$ is the set of points that can be steered to the origin in any finite time, so that

$$\mathscr{C} = \bigcup_{t_1 \geq 0} \mathscr{C}(t_1). \tag{3.4}$$

It is obvious that $\mathscr{C} \subset \mathscr{R}^n$, and a desirable property of the system is that $\mathscr{C} = \mathscr{R}^n$, so that *all* initial states are controllable to the origin. In this case we say that the system is *completely controllable.*

We begin by proving some results for the NLA system.

1. *If* $\mathbf{x}^0 \in \mathscr{C}$ *and if* $\mathbf{y}$ *is a point on the trajectory from* $\mathbf{x}^0$ *to* $\mathbf{0}$, *then* $\mathbf{y} \in \mathscr{C}$. For suppose the trajectory is given by $\mathbf{x}(t)$, with control $\mathbf{u}(t)$, and $\mathbf{y} = \mathbf{x}(\tau)$, $\mathbf{0} = \mathbf{x}(t_1)$. Then, with control $\mathbf{v}(t) = \mathbf{u}(t + \tau)$ and with $\mathbf{x}(0) = \mathbf{y}$, we follow the same trajectory as that from $\mathbf{x}^0$ and reach the origin in time $t_1 - \tau$. Thus, $\mathbf{y} \in \mathscr{C}(t_1 - \tau)$ and $\mathbf{y} \in \mathscr{C}$. This result means that the whole of the successful path from $\mathbf{x}^0$ to $\mathbf{0}$ lies in the controllable set (see Fig. 3.1(a)).

2. $\mathscr{C}$ *is arcwise connected.* If $\mathbf{x}^0$ and $\mathbf{y}^0$ are in $\mathscr{C}$, then, by what we have just proved, there is a path from each point to the origin lying wholly in $\mathscr{C}$. Hence there is an arc in $\mathscr{C}$ connecting $\mathbf{x}^0$ and $\mathbf{y}^0$ (see Fig. 3.1(b)). A similar argument also shows that $\mathscr{C}(t_1)$ is arcwise connected. This proves that $\mathscr{C}$ is not composed of a number of disjoint parts.

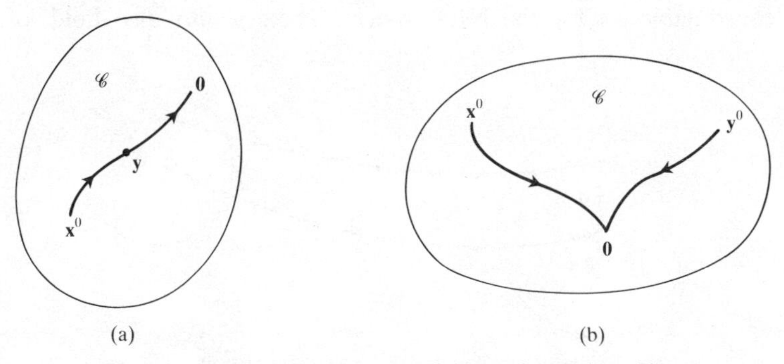

Fig. 3.1. Trajectories in the controllable set for NLA problems.

3. *If* $t_1 < t_2$, $\mathscr{C}(t_1) \subset \mathscr{C}(t_2)$. For suppose that $\mathbf{x}^0$ is any point of $\mathscr{C}(t_1)$, with control $\mathbf{u}(t)$. If we apply the control $\mathbf{v}(t)$ to $\mathbf{x}^0$, where $\mathbf{v} = \mathbf{u}$ for $0 < t < t_1$, and $\mathbf{v} = \mathbf{0}$ for $t_1 < t < t_2$, then $\mathbf{x}(t_1) = \mathbf{0}$ and $\mathbf{x}(t) = \mathbf{0}$ for $t_1 < t < t_2$ since $\mathbf{f}(\mathbf{0}, \mathbf{0}) = \mathbf{0}$. The control $\mathbf{v}$ is integrable and so is admissible, and we have thus shown that the controllable set at any time includes the controllable set at all previous times. Since $\mathbf{0}$ clearly belongs to $\mathscr{C}(0)$, $\mathbf{0} \in \mathscr{C}(t_1)$ and $\mathbf{0} \in \mathscr{C}$.

4. $\mathscr{C}$ *is open if and only if* $\mathbf{0} \in \mathrm{Int}(\mathscr{C})$. Since $\mathbf{0} \in \mathscr{C}$, and $\mathscr{C} = \mathrm{Int}(\mathscr{C})$ when $\mathscr{C}$ is open, it follows at once that $\mathbf{0} \in \mathrm{Int}(\mathscr{C})$. If, on the other hand, $\mathbf{0} \in \mathrm{Int}(\mathscr{C})$, there is a ball $\mathscr{B}(\mathbf{0}, r) \subset \mathscr{C}$. Suppose $\mathbf{u}$ is the control that steers an arbitrary point $\mathbf{x}^0$ to $\mathbf{0}$ in time t_1 (see Fig. 3.2). Since $\mathbf{f}$ is continuously differentiable, the solutions of the NLA equations with this control depend continuously on $\mathbf{x}^0$. Let $\mathbf{y}(t)$ be the solution starting from the point $\mathbf{y}^0 \in \mathscr{B}_0(\mathbf{x}^0, r_0)$. By continuity, $\mathbf{y}(t_1) = \mathbf{y}^1 \in \mathscr{B}(\mathbf{0}, r)$ if we choose r_0 sufficiently small. Since this ball lies in $\mathscr{C}$, it is possible to find a control $\hat{\mathbf{u}}$ that steers $\mathbf{y}^1$ to $\mathbf{0}$ in time t_2, say. Thus it is possible to steer $\mathbf{y}^0$ to $\mathbf{0}$ in time $t_1 + t_2$ by using the control $\mathbf{v}$ defined by

$$\begin{aligned}\mathbf{v}(t) &= \mathbf{u}(t) &&\text{for } 0 < t < t_1,\\ &= \hat{\mathbf{u}}(t - t_1) &&\text{for } t_1 < t < t_1 + t_2.\end{aligned}$$

Hence $\mathbf{y}_0 \in \mathscr{C}(t_1 + t_2)$, and $\mathscr{B}(\mathbf{x}^0, r_0) \subset \mathscr{C}$ for all $\mathbf{x}^0 \in \mathscr{C}$. Thus we have proved that $\mathscr{C}$ is open if $\mathbf{0} \in \mathrm{Int}(\mathscr{C})$.

Since we have to increase the value of t_1 to $t_1 + t_2$ in proving the controllability of $\mathbf{y}^0$, this argument does not prove that $\mathscr{C}(t_1)$ is open. In fact, it is often true that the boundary of $\mathscr{C}(t_1)$ belongs to the set.

This is about as far as we can go in the description of the controllable set for the NLA system. These results also hold, of

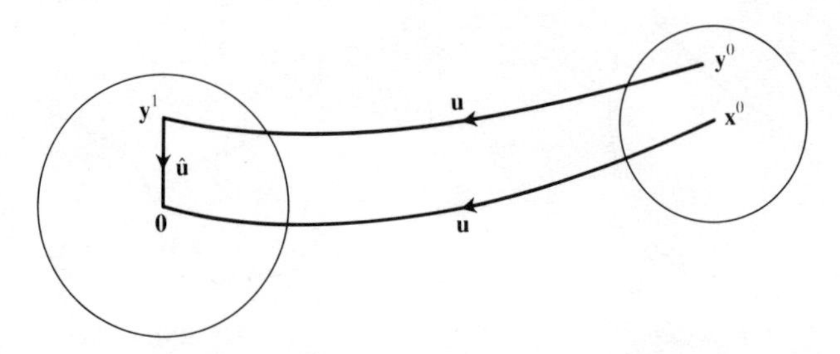

Fig. 3.2. The origin is an internal point of the open controllable set.

course, for the LA system, but we can proceed much further for the linear system. Our future concentration on the linear system is not made simply to avoid the difficulties of nonlinearity. For general systems, small variations from a desired path are governed by a linearized form of the equations and it is desirable to know if such small variations are controllable. If so, the correct path can be rejoined.

3.2. Controllability for the linear system

The state equations for the linear autonomous system are given by (3.3) and we can at once write down their solution. From (2.34), we see immediately that

$$\mathbf{x}(t) = \exp(\mathbf{A}t)\left(\mathbf{x}^0 + \int_0^t \exp(-\mathbf{A}\tau)\mathbf{B}\mathbf{u}(\tau)\,\mathrm{d}\tau\right). \tag{3.5}$$

It follows that $\mathbf{x}^0 \in \mathscr{C}(t_1)$ if and only if there is a control $\mathbf{u}(t) \in \mathscr{U}$, the chosen set of admissible controls, such that

$$\mathbf{x}^0 = -\int_0^{t_1} \exp(-\mathbf{A}\tau)\mathbf{B}\mathbf{u}(\tau)\,\mathrm{d}\tau. \tag{3.6}$$

We now prove some results about the controllable set. Our goal is to find conditions that ensure that all points are controllable, that is, $\mathscr{C} = \mathscr{R}^n$. The control sets that we shall use are integrable, with or without bounds: $\mathscr{U}$ can be taken to be either $\mathscr{U}_{\mathrm{u}}$ or $\mathscr{U}_{\mathrm{b}}$.

1. *$\mathscr{C}(t_1)$ is symmetric and convex.* If $\mathbf{x}^0 \in \mathscr{C}(t_1)$ with control $\mathbf{u}(t)$, it follows from (3.6) that $-\mathbf{x}^0 \in \mathscr{C}(t_1)$ with control $-\mathbf{u}(t)$, and $\mathscr{C}(t_1)$ is symmetric.

The set $\mathscr{U}$ is convex, since if $\mathbf{u}^1$ and $\mathbf{u}^2$ belong to $\mathscr{U}$, $c\mathbf{u}^1 + (1-c)\mathbf{u}^2 \in \mathscr{U}$ for $0 < c < 1$. If these are the successful controls for two points $\mathbf{x}^1$ and $\mathbf{x}^2$ belonging to $\mathscr{C}(t_1)$, it follows from (3.6) that

$$c\mathbf{x}^1 + (1-c)\mathbf{x}^2 = -\int_0^{t_1} \exp(-\mathbf{A}\tau)\mathbf{B}[c\mathbf{u}^1 + (1-c)\mathbf{u}^2]\,\mathrm{d}\tau, \tag{3.7}$$

and the line segment joining $\mathbf{x}^1$ and $\mathbf{x}^2$ belongs to $\mathscr{C}(t_1)$ which is therefore convex.

2. *$\mathscr{C}$ is symmetric and convex.* The symmetry of $\mathscr{C}$ is obvious, but the convexity does not follow at once from §3.2(1) above because the union of convex sets is not necessarily convex. However, if $\mathbf{x}^1 \in \mathscr{C}(t_1)$

and $\mathbf{x}^2 \in \mathscr{C}(t_2)$ with $t_2 > t_1$, then $\mathbf{x}^1 \in \mathscr{C}(t_2)$ since $\mathscr{C}(t_1) \subset \mathscr{C}(t_2)$ from §3.1(3). Therefore, since $\mathscr{C}(t_2)$ is convex, $c\mathbf{x}^1 + (1-c)\mathbf{x}^2 \in \mathscr{C}(t_2) \subset \mathscr{C}$ for $0 < c < 1$ and $\mathscr{C}$ is convex.

As an example of these results, consider an unstable system with two components. Any initial deviation from the origin will, if uncontrolled, lead to a growing departure from the equilibrium state. For a controlled system with these characteristics we can consider the state equations

$$\dot{x}_1 = x_1 + u_1, \quad \dot{x}_2 = x_2 + u_1, \tag{3.8}$$

and $u_1 \in \mathscr{U}_b$, that is, $-1 \leq u_1 \leq 1$. In this example, $n = 2$ and $m = 1$ and the matrices $\mathbf{A}$ and $\mathbf{B}$ are given by

$$\mathbf{A} = \begin{bmatrix} 1 & 0 \\ 0 & 1 \end{bmatrix}, \quad \mathbf{B} = \begin{bmatrix} 1 \\ 1 \end{bmatrix}. \tag{3.9}$$

From the general result given by (3.6) we see that $\mathbf{x} \in \mathscr{C}(t_1)$ if

$$x_1 = -\int_0^{t_1} \exp(-\tau) u_1 \, d\tau = x_2, \tag{3.10}$$

since $\exp(\mathbf{A}t) = \exp(t)\mathbf{I}$. Because $|u_1| \leq 1$,

$$|x_1| \leq \int_0^{t_1} \exp(-\tau) 1 \, d\tau = 1 - \exp(-t_1), \tag{3.11}$$

and equality is possible. Hence $\mathscr{C}(t_1)$ is a closed line interval in $\mathscr{R}^2$,

$$\mathscr{C}(t_1) = \{x_1 = x_2, |x_1| \leq 1 - \exp(-t_1)\}, \tag{3.12}$$

and $\mathscr{C}$ is the open interval

$$\mathscr{C} = \{x_1 = x_2, |x_1| < 1\}. \tag{3.13}$$

As part of $\mathscr{R}^2$, $\mathrm{Int}(\mathscr{C})$ is empty and $\mathscr{C}$ is not open. It is impossible to control initial states that do not lie on this interval, however close to the origin they may be. Thus, although it is possible to control either component independently when the initial deviation from the origin is less than one in magnitude, it is, in general, impossible to control both components simultaneously with identical controls. Only if the initial deviations of the two components are equal can this be done.

In general, two requirements are needed if the system is to be completely controllable. The controllable set must be n-dimensional and not collapse on to a hyperplane as it does in this example. It must also be unbounded, even when the controls are restricted to the bounded set $\mathscr{U}_b$.

3.3. The controllability matrix

In order to derive a condition that enables us to settle the question of complete controllability, we introduce the *controllability matrix* $\mathbf{M}$. This is a matrix of n rows and mn columns, formed from the $n \times m$ matrix $\mathbf{B}$ and its products with powers of the $n \times n$ matrix $\mathbf{A}$. We define $\mathbf{M}$ by

$$\mathbf{M} = [\mathbf{B} \;\; \mathbf{AB} \;\; \mathbf{A}^2\mathbf{B} \;\ldots\; \mathbf{A}^{n-1}\mathbf{B}]. \tag{3.14}$$

The results we establish depend on the rank of this matrix and the eigenvalues of $\mathbf{A}$. It is clear that the system cannot be completely controllable when $\mathbf{0} \notin \operatorname{Int}(\mathscr{C})$, for then there are points close to the origin that are not controllable. The first result relates to this requirement.

1. $\mathbf{0} \in \operatorname{Int}(\mathscr{C})$ *if and only if rank* $\mathbf{M} = n$. In the example discussed in §3.2, it is easy to show that

$$\mathbf{M} = \begin{bmatrix} 1 & 1 \\ 1 & 1 \end{bmatrix}, \tag{3.15}$$

which has rank equal to one. For the positioning problem in §1.4, the matrices are

$$\mathbf{A} = \begin{bmatrix} 0 & 1 \\ 0 & 0 \end{bmatrix}, \quad \mathbf{B} = \begin{bmatrix} 0 \\ 1 \end{bmatrix}, \quad \mathbf{M} = \begin{bmatrix} 0 & 1 \\ 1 & 0 \end{bmatrix}, \tag{3.16}$$

and the rank of $\mathbf{M}$ is equal to two. We have already established that, in the first example, there are points near the origin that are not controllable, while an extension of the arguments used in §1.4 shows that all points are controllable. This is consistent with the criterion based on the rank of $\mathbf{M}$ which we must now establish.

Suppose first that rank $\mathbf{M} < n$. Then there is at least one vector $\mathbf{y} \in \mathscr{R}^n$, with $\|\mathbf{y}\| = 1$, that is orthogonal to every column of $\mathbf{M}$. This implies that

$$\mathbf{y}^{\mathrm{T}}\mathbf{B} = \mathbf{0}, \quad \mathbf{y}^{\mathrm{T}}\mathbf{A}^k\mathbf{B} = \mathbf{0}, \quad k = 1, 2, \ldots, n-1. \tag{3.17}$$

The Cayley–Hamilton theorem (see §2.5) then shows that

$$\mathbf{y}^{\mathrm{T}}\mathbf{A}^k\mathbf{B} = \mathbf{0}, \quad \text{for all integers } k, \tag{3.18}$$

and from (2.33) we see that

$$\mathbf{y}^{\mathrm{T}}\exp(-\mathbf{A}\tau)\mathbf{B} = \mathbf{0}. \tag{3.19}$$

Hence, for all $\mathbf{x}^0 \in \mathscr{C}(t_1)$,

$$\mathbf{y}^{\mathrm{T}}\mathbf{x}^0 = -\int_0^{t_1} \mathbf{y}^{\mathrm{T}} \exp(-\mathbf{A}\tau)\mathbf{B}\mathbf{u}(\tau)\,\mathrm{d}\tau = 0. \tag{3.20}$$

Therefore $\mathscr{C}(t_1)$ lies in a hyperplane with normal $\mathbf{y}$ for all $t_1 > 0$ and $\mathscr{C}$ lies in the same hyperplane. Thus $\mathbf{0} \notin \operatorname{Int} \mathscr{C}$.

Now suppose that $\mathbf{0} \notin \operatorname{Int} \mathscr{C}$. Since $\mathscr{C}(t_1) \subset \mathscr{C}$, $\mathbf{0} \notin \operatorname{Int} \mathscr{C}(t_1)$ for any t_1, but $\mathbf{0} \in \mathscr{C}(t_1)$ and $\mathscr{C}(t_1)$ is convex. For each value of t_1, there is a hyperplane through $\mathbf{0}$ supporting $\mathscr{C}(t_1)$. If the normal to this hyperplane is $\mathbf{b}(t_1)$, then for all $\mathbf{x}^0 \in \mathscr{C}(t_1)$, $\mathbf{b}^{\mathrm{T}}\mathbf{x}^0 \le 0$ and

$$\int_0^{t_1} \mathbf{b}^{\mathrm{T}} \exp(-\mathbf{A}\tau)\mathbf{B}\mathbf{u}\,\mathrm{d}\tau = -\mathbf{b}^{\mathrm{T}}\mathbf{x}^0 \ge 0 \tag{3.21}$$

for all $\mathbf{u} \in \mathscr{U}$. Since $-\mathbf{u} \in \mathscr{U}$ also, we must have equality, and the vanishing of the integral implies that

$$\mathbf{b}^{\mathrm{T}}\exp(-\mathbf{A}\tau)\mathbf{B} = \mathbf{0} \quad \text{for } 0 \le \tau \le t_1. \tag{3.22}$$

If we put $\tau = 0$, we obtain $\mathbf{b}^{\mathrm{T}}\mathbf{B} = \mathbf{0}$. If we differentiate (3.22) k times and set $\tau = 0$, we obtain $\mathbf{b}^{\mathrm{T}}\mathbf{A}^k\mathbf{B} = \mathbf{0}$. It follows that $\mathbf{b}$ is orthogonal to all the columns of $\mathbf{M}$, and the rank of $\mathbf{M}$ is less than n.

This establishes the criterion relating the rank of the controllability matrix to the origin being an interior point of the controllable set and hence to the set being open, by §3.2(3). If the rank of $\mathbf{M}$ is less than n, $\mathscr{C}$ lies in a hyperplane in $\mathscr{R}^n$ and the system is definitely not completely controllable. However, if the rank is equal to n, the system may or may not be completely controllable. To settle this question, we have now to distinguish between our two control sets, $\mathscr{U}_{\mathrm{u}}$ and $\mathscr{U}_{\mathrm{b}}$.

2. *If rank* $\mathbf{M} = n$, *and* $\mathbf{u} \in \mathscr{U}_{\mathrm{u}}$, $\mathscr{C} = \mathscr{R}^n$. We have established that $\mathbf{0}$ is an interior point of $\mathscr{C}$, so there is a ball $\mathscr{B}(\mathbf{0}, r) \in \mathscr{C}$. Let $\mathbf{x}^0$ be an arbitrary point in $\mathscr{R}^n$. Then $\mathbf{y}^0 = c\mathbf{x}^0$, where $c = \frac{1}{2}r/\|\mathbf{x}^0\|$, is in the ball and hence is controllable with a control $\mathbf{v}$, say, belonging to $\mathscr{U}_{\mathrm{u}}$. Thus,

$$\mathbf{y}^0 = -\int_0^{t_1} \exp(-\mathbf{A}\tau)\mathbf{B}\mathbf{v}(\tau)\,\mathrm{d}\tau, \tag{3.23}$$

and so

$$\mathbf{x}^0 = -\int_0^{t_1} \exp(-\mathbf{A}\tau)\mathbf{B}\mathbf{u}(\tau)\,\mathrm{d}\tau, \tag{3.24}$$

with the control $\mathbf{u} = \mathbf{v}/c$, which belongs to $\mathscr{U}_{\mathrm{u}}$. Hence, the arbitrary point $\mathbf{x}^0 \in \mathscr{C}$. A similar argument shows that $\mathscr{C}(t_1) = \mathscr{R}^n$ for all $t_1 > 0$.

In the rest of this section, the control set will be $\mathscr{U}_{\mathrm{b}}$.

3. *If rank* $\mathbf{M} = n$, *and* $\operatorname{Re} \lambda_i < 0$ *for each eigenvalue* λ_i *of* $\mathbf{A}$, $\mathscr{C} = \mathscr{R}^n$. Choose $\mathbf{x}^0 \in \mathscr{R}^n$, and make a similarity transformation $\mathbf{x} = \mathbf{P}\mathbf{y}$ to reduce the equation $\dot{\mathbf{x}} = \mathbf{A}\mathbf{x}$ to $\dot{\mathbf{y}} = \mathbf{\Lambda}\mathbf{y}$, where $\mathbf{\Lambda}$ is the diagonal matrix containing the eigenvalues. If we set $\mathbf{u} = \mathbf{0}$, the solution is given by

$$\mathbf{y}(t) = \mathbf{\Phi}(t)\mathbf{y}^0, \tag{3.25}$$

where $\mathbf{\Phi}(t)$ is the diagonal matrix with elements equal to $\exp(\lambda_i t)$ and $\mathbf{y}^0 = \mathbf{P}^{-1}\mathbf{x}^0$. (See Exercise 2.3(e).) It follows that $\|\mathbf{y}\| \to 0$ as $t \to \infty$, and hence that $\|\mathbf{x}(t_1)\| < \delta$ for an arbitrarily small positive δ at some finite time t_1. We know that there is a ball $\mathscr{B}(\mathbf{0}, \delta) \subset \mathscr{C}$, because $\mathbf{0}$ is an interior point of $\mathscr{C}$, and so $\mathbf{x}(t_1)$ can be steered to $\mathbf{0}$ in some time t_2 by a control $\mathbf{v}$, say, belonging to $\mathscr{U}_{\mathrm{b}}$. Piecing together the zero control and $\mathbf{v}$, we have an integrable bounded control and $\mathbf{x}^0$ can be steered to $\mathbf{0}$ in time $t_1 + t_2$. Since $\mathbf{x}^0$ was an arbitrary point in $\mathscr{R}^n$, we have shown that $\mathscr{C} = \mathscr{R}^n$. Note that we could not deduce this result directly from (3.25), since this only shows that the origin can be reached in an infinite time, and the definition of the controllable set requires the target to be reached in some *finite* time. The above argument needs amending if the eigenvalues of $\mathbf{A}$ are not all distinct. The similarity transformation leaves some non-zero elements above the main diagonal, but it is still true that the solution for $\mathbf{y}$ tends to $\mathbf{0}$. (Compare the solution of Exercise 2.3(b).)

In a number of common examples, this result is not quite strong enough to provide the criterion we require, since some of the eigenvalues are purely imaginary.

4. *If rank* $\mathbf{M} = n$, *and* $\operatorname{Re} \lambda_i \le 0$ *for each eigenvalue* λ_i *of* $\mathbf{A}$, *then* $\mathscr{C} = \mathscr{R}^n$. Suppose that $\mathscr{C} \ne \mathscr{R}^n$ and that $\mathbf{y} \notin \mathscr{C}$. Then there is a hyperplane with normal $\mathbf{b}$ separating $\mathbf{y}$ and $\mathscr{C}$, such that

$$\mathbf{b}^{\mathrm{T}}\mathbf{x}^0 \le p \quad \text{for all } \mathbf{x}^0 \in \mathscr{C} \quad \text{and} \quad \mathbf{b}^{\mathrm{T}}\mathbf{y} > p. \tag{3.26}$$

We need to show that

$$\mathbf{b}^{\mathrm{T}}\mathbf{x}^0 = -\int_0^{t_1} \mathbf{b}^{\mathrm{T}} \exp(-\mathbf{A}\tau)\mathbf{B}\mathbf{u}(\tau)\,\mathrm{d}\tau > p \tag{3.27}$$

for t_1 sufficiently large and for some control $\mathbf{u} \in \mathscr{U}_{\mathrm{b}}$, which will be a contradiction. Define $\mathbf{z}(t) = \mathbf{b}^{\mathrm{T}} \exp(-\mathbf{A}\tau)\mathbf{B}$. Because rank $\mathbf{M} = n$, $\mathbf{z}(t) \neq \mathbf{0}$ for $0 \leq t \leq t_1$, and we can choose $u_i(t) = -\operatorname{sgn} z_i(t)$ so that

$$\mathbf{b}^{\mathrm{T}}\mathbf{x}^0 = \int_0^{t_1} |\mathbf{v}(t)|\,\mathrm{d}t. \tag{3.28}$$

Each component of $\mathbf{v}$ is a combination of terms of the general form $q(t)\exp(-\lambda_i t)$, where q is a polynominal and λ_i is an eigenvalue of $\mathbf{A}$. It is clear that the right-hand side of (3.28) can be made arbitrarily large for large t_1 if the exponentially growing terms with Re $\lambda_i < 0$ are present. The terms corresponding to the eigenvalues that have Re $\lambda_i = 0$ are either polynomials in t or are periodic in t. The polynomial terms, and the periodic terms over one period, both make a positive contribution to the right-hand side of (3.28) over a non-zero interval, so that it can be made arbitrarily large by choosing t_1 large enough. We have thus obtained the required contradiction, and $\mathscr{C} = \mathscr{R}^n$.

Finally, we have a criterion for the system not to be completely controllable. We have already seen that this is so if rank $\mathbf{M}$ is less than n.

5. *If rank* $\mathbf{M} = n$, *and* $\mathbf{A}$ *has an eigenvalue with positive real part,* $\mathscr{C} \neq \mathscr{R}^n$. Suppose λ is an eigenvalue of $\mathbf{A}$ and Re $\lambda > 0$, and let $\mathbf{y}$ be the associated left eigenvector, so that $\mathbf{y}^{\mathrm{T}}\mathbf{A} = \lambda\mathbf{y}^{\mathrm{T}}$. Then $\mathbf{y}^{\mathrm{T}}\mathbf{A}^k = \lambda^k\mathbf{y}^{\mathrm{T}}$ so that

$$\mathbf{y}^{\mathrm{T}}\exp(-\mathbf{A}\tau) = \exp(-\lambda\tau)\mathbf{y}^{\mathrm{T}} \tag{3.29}$$

and

$$\mathbf{y}^{\mathrm{T}}\mathbf{x}^0 = -\int_0^{t_1} \exp(-\lambda\tau)\mathbf{y}^{\mathrm{T}}\mathbf{B}\mathbf{u}(\tau)\,\mathrm{d}\tau. \tag{3.30}$$

Because $\mathbf{u}$ belongs to the bounded set $\mathscr{U}_{\mathrm{b}}$, this integral is bounded by p, say, as $t_1 \to \infty$, and so $\mathbf{y}^{\mathrm{T}}\mathbf{x}^0 < p$. The controllable points lie in a half-space on one side of a hyperplane in $\mathscr{R}^n$ and so $\mathscr{C} \neq \mathscr{R}^n$.

3.4. The reachable set

So far we have only considered controllability to $\mathbf{0}$. We can define $\mathscr{C}(t_1, \mathbf{x}^1)$ as the set of points that are controllable to $\mathbf{x}^1$ in time t_1 in a similar way, and $\mathbf{x}^0$ will belong to this set if

$$\mathbf{x}^0 = \exp(-\mathbf{A}t_1)\mathbf{x}^1 - \int_0^{t_1} \exp(-\mathbf{A}\tau)\mathbf{B}\mathbf{u}(\tau)\,\mathrm{d}\tau \tag{3.31}$$

for some control $\mathbf{u} \in \mathscr{U}$. This result follows at once from the general solution of the linear system given by (3.5). Many results that we have proved for $\mathscr{C}(t_1)$ are also true for $\mathscr{C}(t_1, \mathbf{x}^1)$, notably its convexity, but it may no longer be possible to hold the state at the target point with an admissible control. If there is no control $\mathbf{u}$ such that $\mathbf{A}\mathbf{x}^1 + \mathbf{B}\mathbf{u} = \mathbf{0}$ we cannot deduce that $\mathscr{C}(t_1, \mathbf{x}^1) \subset \mathscr{C}(t_2, \mathbf{x}^1)$ for $t_2 > t_1$, for example.

We can define the *reachable set* $\mathscr{R}(t_1, \mathbf{x}^0)$ as the set of points that are reachable from the initial point $\mathbf{x}^0$ in time t_1. From (3.5), $\mathbf{x}^1 \in \mathscr{R}(t_1, \mathbf{x}^0)$ when

$$\mathbf{x}^1 = \exp(\mathbf{A}t_1)\left(\mathbf{x}^0 + \int_0^{t_1} \exp(-\mathbf{A}\tau)\mathbf{B}\mathbf{u}(\tau)\,\mathrm{d}\tau\right) \tag{3.32}$$

for some $\mathbf{u} \in \mathscr{U}$. There is obviously a reciprocal relationship between the two sets: if $\mathbf{x}^1 \in \mathscr{R}(t_1, \mathbf{x}^0)$, then $\mathbf{x}^0 \in \mathscr{C}(t_1, \mathbf{x}^1)$. Also, we can define the *time-reversed system* as the system with the state equation (3.3) replaced by

$$\dot{\mathbf{x}} = -\mathbf{A}\mathbf{x} - \mathbf{B}\mathbf{u}. \tag{3.33}$$

If we change τ to $t_1 - \tau$ it is easy to see, from (3.31) and (3.32), that the controllable set for the time-reversed system is equal to the reachable set for the original system. Thus the general properties we have found to hold for the controllable set will also hold for the reachable set, since (3.33) is an LA system. The same reciprocity holds for the NLA system, but not for non-autonomous systems.

To illustrate these ideas, let us consider a very simple one-dimensional system,

$$\dot{x}_1 = x_1 + u_1, \quad x_1^0 = \tfrac{1}{2}, \quad |u_1| \leq 1. \tag{3.34}$$

Points in the reachable set at time t_1 are given by

$$x_1^1 = \exp(t_1)\left(\tfrac{1}{2} + \int_0^{t_1} \exp(-\tau)u_1(\tau)\,d\tau\right) \tag{3.35}$$

and the bounds on the possible values u_1 can take show that $\mathscr{R}(t_1, \frac{1}{2})$ is the closed interval from $1 - \frac{1}{2}\exp(t_1)$ to $\frac{3}{2}\exp(t_1) - 1$. Similarly, points in the reachable set for the time-reversed system are given by

$$x_1^1 = \exp(-t_1)\left(\tfrac{1}{2} - \int_0^{t_1} \exp(\tau)u_1(\tau)\,d\tau\right), \tag{3.36}$$

so this reachable set is the closed interval from $\frac{3}{2}\exp(-t_1) - 1$ to $1 - \frac{1}{2}\exp(-t_1)$. This set is also the set controllable to the point $\frac{1}{2}$ for the system (3.34). These sets are sketched in Fig. 3.3 and show that all points in $\mathscr{R}^1$ are reachable from $\frac{1}{2}$ in finite time, but only those points belonging to the open interval $-1 < x_1^0 < 1$ are controllable to $\frac{1}{2}$.

An important point illustrated by these particular results is that the reachable and controllable sets are continuous in t_1. To show that this is a general property of the LA system, let us consider, for simplicity, the set reachable from the origin with bounded controls, which we shall denote by $\mathscr{R}(t)$. The points $\mathbf{x} \in \mathscr{R}(t)$ are given by

$$\mathbf{x} = \int_0^t \exp(\mathbf{A}\tau)\mathbf{B}\mathbf{u}(\tau)\,d\tau \tag{3.37}$$

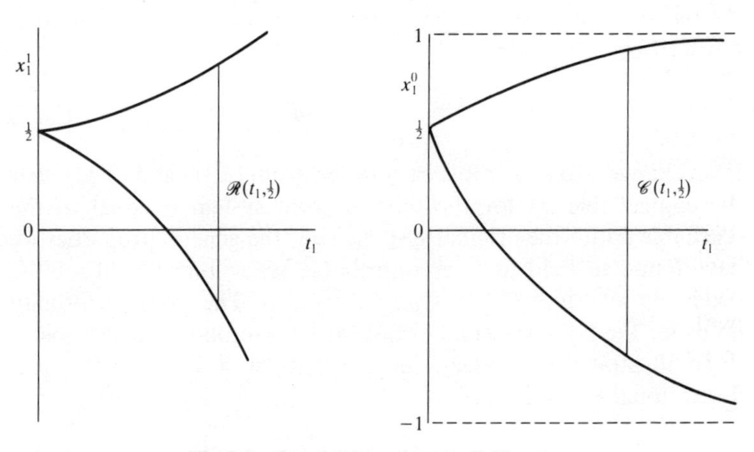

Fig. 3.3. Reachable and controllable sets.

with $\mathbf{u} \in \mathscr{U}_b$. The corresponding results for an arbitrary initial point, and also for the controllable set, follow by similar arguments. In order to establish the continuity of $\mathscr{R}(t_1)$, we have to show that, for all t_2 such that $|t_2 - t_1| < \delta(\varepsilon)$, there is a point $\mathbf{y} \in \mathscr{R}(t_2)$ within a distance ε from an arbitrary point $\mathbf{x} \in \mathscr{R}(t_1)$ and a point $\mathbf{x} \in \mathscr{R}(t_1)$ within a distance ε from an arbitrary point $\mathbf{y} \in \mathscr{R}(t_2)$. We use the norm defined in (2.46), and we suppose that t_1 and t_2 lie between 0 and T. Let

$$M = \max_{0 < \tau < T} |\exp(\mathbf{A}\tau)\mathbf{B}|. \tag{3.38}$$

Suppose that $\hat{\mathbf{u}}$ is the control for an arbitrarily chosen point $\hat{\mathbf{x}}$ in $\mathscr{R}(t_1)$. Let $t_3 = \min(t_1, t_2)$ and define the control $\hat{\mathbf{v}}$ so that

$$\hat{\mathbf{v}}(t) = \hat{\mathbf{u}}(t) \quad \text{for } 0 \le t \le t_3, \quad \hat{\mathbf{v}}(t) = \mathbf{0} \quad \text{for } t_3 < t \le T. \tag{3.39}$$

If

$$\hat{\mathbf{y}} = \int_0^{t_2} \exp(\mathbf{A}\tau)\mathbf{B}\hat{\mathbf{v}}\, \mathrm{d}\tau, \tag{3.40}$$

then $\hat{\mathbf{y}} \in \mathscr{R}(t_2)$ and

$$|\hat{\mathbf{y}} - \hat{\mathbf{x}}| = \left| \int_{t_3}^{t_2} \exp(\mathbf{A}\tau)\mathbf{B}\hat{\mathbf{v}}\, \mathrm{d}\tau \right| \le M|t_2 - t_3| < \varepsilon \tag{3.41}$$

if $\delta < \varepsilon/M$, so there is a point in $\mathscr{R}(t_2)$ arbitrarily close to every point in $\mathscr{R}(t_1)$. A similar proof holds with $\mathscr{R}(t_1)$ and $\mathscr{R}(t_2)$ interchanged, so we have proved that $\mathscr{R}(t)$ is continuous.

It is also possible to prove that the reachable set is *compact*. This means that the set is bounded, which is obviously true, and closed. We have to show that any converging sequence of points in the reachable set has a limit in that set, so that there is a corresponding sequence of controls with a limit within the admissible set of controls. To prove this we first have to define a suitable norm for the control functions $\mathbf{u}(t)$ over the interval $(0, t_1)$ since we need to work with limits of sequences of functions. The L^2 norm for one component $u_i(t)$ is defined by

$$L^2[u_i] = \left| \int_0^{t_1} |u_i(\tau)|^2\, \mathrm{d}\tau \right|^{1/2}. \tag{3.42}$$

With $\mathbf{u} \in \mathcal{U}_b$, $L^2[u_i] \le t_1^{1/2}$. These controls are compact in the sense that, if $u_i^{(k)}$ is a sequence of controls such that

$$\lim_{k\to\infty} \int_0^{t_1} f(\tau)u_i^{(k)}\, d\tau = \int_0^{t_1} f(\tau)u_i\, d\tau, \tag{3.43}$$

for all integrable functions $f(t)$, then u_i is integrable and $|u_i| \le 1$. For suppose that $u_i > 1$ over part of the interval $[0, t_1]$ of measure h. Then if we define $f(t)$ to be equal to one when $u_i > 1$ and zero elsewhere, the integral on the right-hand side of (3.43) is greater than h. But the integral on the left is less than or equal to h, since $u_i^{(k)} \le 1$. This contradiction, and a similar one for $u_i < -1$, applied to all components of $\mathbf{u}$ is enough to prove compactness (technically this is known as weak sequential compactness). Since the points in the reachable set are defined in terms of integrals of the same form as (3.43) for specific functions $f(t)$, the reachable set is closed.

A final property of the reachable set is contained in what is known as the *bang–bang principle*. There is some confusion about the meaning of this principle. It is sometimes taken to mean that, in time-optimal control problems with bounded controls, the optimal solution is reached by the use of bang–bang controls, that is, controls in which every component takes only its extreme values, switching discontinously between them. This is indeed true, as we shall find in the next chapter. But the bang–bang principle actually states that all points that are reachable by using bounded controls are reachable using bang–bang controls. In other words,

$$\mathscr{R}(t_1, \mathbf{x}^0) \text{ with } \mathbf{u} \in \mathcal{U}_b = \mathscr{R}(t_1, \mathbf{x}^0) \text{ with } \mathbf{u} \in \mathcal{U}_{bb}. \tag{3.44}$$

A proof of this principle for LA problems will be postponed to Chapter 4, because it makes use of some results derived in that chapter.

This concludes our discussion of the controllable and reachable sets in general, which provides the essential basis for the solution of the time-optimal control problem. We have made no restriction on the number of controls used, that is, on the dimension of the control vector $\mathbf{u}$. There is a possible redundancy among these controls, which it is sometimes advantageous to remove. This can be done without losing anything of practical importance in applications of the theory.

3.5. Redundant controls

In the state equation (3.3) of the linear autonomous problem, the controls appear in the term $\mathbf{Bu}$ only, where $\mathbf{B}$ is an $n \times m$ matrix, and $\mathbf{u}$ has m components. Since the solution of (3.3) from a given initial point is unique, different controls will lead to different trajectories or responses provided $\mathbf{Bu} = \mathbf{Bv}$ implies that $\mathbf{u} = \mathbf{v}$, that is, provided $\mathbf{B}$ has m linearly independent columns. If $m > n$ this is impossible, and it is also clear that the number of independent controls can be reduced. Suppose that $\mathbf{B}$ has $\hat{m}$ linearly independent columns, with $\hat{m} \leq n$. If $\hat{m} < m$ we can find an elementary matrix $\mathbf{P}$ for column operations such that

$$\mathbf{u} = \mathbf{P}\hat{\mathbf{u}}, \quad \mathbf{BP} = \hat{\mathbf{B}}, \quad \mathbf{Bu} = \hat{\mathbf{B}}\hat{\mathbf{u}}, \tag{3.45}$$

where $\hat{\mathbf{B}}$ contains $\hat{m}$ independent columns of $\mathbf{B}$ and $m - \hat{m}$ columns of zeros. We can rearrange the columns so that the non-zero columns occupy the first $\hat{m}$ positions. If we then define $\mathbf{C}$ as the $n \times \hat{m}$ matrix containing the non-zero columns of $\hat{\mathbf{B}}$ and $\mathbf{v}$ as the $\hat{m}$-dimensional vector containing the first $\hat{m}$ components of $\hat{\mathbf{u}}$,

$$\mathbf{Cv} = \hat{\mathbf{B}}\hat{\mathbf{u}} = \mathbf{Bu}, \tag{3.46}$$

and we have a system with $\hat{m}$ independent controls and an associated matrix with linearly independent columns.

In general, then, we can confine our attention to systems of the standard form (3.3), but with $m \leq n$ and with rank $\mathbf{B} = m$. This system has no redundant controls. To show how redundancy can be removed, consider an example in which $m = 3$ and $n = 2$, and

$$\mathbf{B} = \begin{bmatrix} 1 & 2 & 3 \\ 1 & 2 & 3 \end{bmatrix}, \quad \mathbf{u}^{\mathrm{T}} = [u_1 \quad u_2 \quad u_3]. \tag{3.47}$$

The elementary matrix

$$\mathbf{P} = \begin{bmatrix} 1 & -2 & -3 \\ 0 & 1 & 0 \\ 0 & 0 & 1 \end{bmatrix} \tag{3.48}$$

changes the matrix to

$$\hat{\mathbf{B}} = \begin{bmatrix} 1 & 0 & 0 \\ 1 & 0 & 0 \end{bmatrix}, \tag{3.49}$$

with $\mathbf{u} = \mathbf{P}\hat{\mathbf{u}}$, so we can redefine the system with a single control v_1 and a 2×1 matrix $\mathbf{C}$, where

$$\mathbf{C} = \begin{bmatrix} 1 \\ 1 \end{bmatrix}, \quad \text{and} \quad v_1 = u_1 + 2u_2 + 3u_3. \tag{3.50}$$

Any choices of the original controls that leave the quantity v_1 unchanged do not affect the trajectory, so the original system contained a double redundancy.

Although most of the results concerning the time-optimal solution hold whether or not the redundant controls have been eliminated, there is one place where, as we shall see, it is important that we have a system without redundancy. In the next chapter we solve the time-optimal control problem.

Exercises 3

(*Solutions on p. 219*)

3.1 If $\dot{\mathbf{x}} = \mathbf{A}\mathbf{x} + \mathbf{B}\mathbf{u}$, which of the following systems are completely controllable (i) with $\mathbf{u} \in \mathscr{U}_u$, (ii) with $\mathbf{u} \in \mathscr{U}_b$?

(a) $\mathbf{A} = \begin{bmatrix} 0 & 1 \\ 0 & 1 \end{bmatrix}, \quad \mathbf{B} = \begin{bmatrix} 1 & -1 \\ 0 & 0 \end{bmatrix}.$

(b) $\mathbf{A} = \begin{bmatrix} 0 & 0 \\ 1 & 1 \end{bmatrix}, \quad \mathbf{B} = \begin{bmatrix} 1 & -1 \\ 0 & 0 \end{bmatrix}.$

(c) $\mathbf{A} = \begin{bmatrix} p & 1 & 0 \\ 0 & p & 1 \\ 0 & 0 & p \end{bmatrix}, \quad \mathbf{B} = \begin{bmatrix} q \\ r \\ s \end{bmatrix},$

where p, q, r, s are real numbers.

3.2 For the one-dimensional system with $\dot{x}_1 = -x_1 + u_1$ and $|u_1(t)| \le 1$, find $\mathscr{R}(t_1, 0)$, $\mathscr{R}(t_1, 2)$, $\mathscr{C}(t_1, 0)$, and $\mathscr{C}(t_1, 2)$. (These sets are defined in §3.4.) If $\mathscr{C}(a) = \cup_{t_1 \ge 0} \mathscr{C}(t_1, a)$ and $\mathscr{R}(a) = \cup_{t_1 \ge 0} \mathscr{R}(t_1, a)$, find $\mathscr{C}(0)$, $\mathscr{C}(2)$, $\mathscr{R}(0)$, $\mathscr{R}(2)$.

Show that $\mathscr{R}(t_1, 2) \not\subset \mathscr{R}(t_2, 2)$ when $t_1 < t_2$.

3.3 Verify the reciprocity between the reachable and controllable sets by repeating Exercise 3.2 for the time-reversed system $\dot{x}_1 = x_1 - u_1$.

3.4 Find $\mathscr{C}(t_1, \mathbf{0})$ and $\mathscr{C}(\mathbf{0})$ for the two-dimensional system with a single bounded control u_1 when

$$\mathbf{A} = \begin{bmatrix} 0 & 1 \\ 0 & 0 \end{bmatrix}, \quad \mathbf{B} = \begin{bmatrix} 1 \\ 0 \end{bmatrix}.$$

Find also $\mathscr{R}(t_1, \mathbf{x}^0)$ when $\mathbf{x}^0 = (\mathbf{p}, \mathbf{q})^{\mathrm{T}}$.

3.5 Consider the system $\dot{\mathbf{x}} = \mathbf{u}$, with $m = n = 2$ and $\mathbf{x}^0 = (-1, 0)^{\mathrm{T}}$. The controls are bounded, so that $|u_i| \leq 1$ for $i = 1, 2$. Show that $\mathscr{R}(t_1, \mathbf{x}^0)$ is a square, and show that the origin first belongs to this square when $t_1 = 1$. Suppose that is desired to steer this point $\mathbf{x}^0$ to $\mathbf{x}^1 = (0, p)^{\mathrm{T}}$ in unit time. Find the controls needed to do this when $p = 1$. If $-1 < p < 1$, show that the controls are not uniquely determined.

3.6 Consider the system with bounded control u_1 and

$$\mathbf{A} = \begin{bmatrix} 0 & 1 \\ -1 & 0 \end{bmatrix}, \quad \mathbf{B} = \begin{bmatrix} 0 \\ 1 \end{bmatrix}.$$

If the initial point is $\mathbf{x}^0 = (r \cos \alpha, r \sin \alpha)^{\mathrm{T}}$, show that $\mathbf{x}^0 \in \mathscr{C}(t_1, \mathbf{0})$ if a bounded control u_1 exists such that

$$r \cos \alpha = \int_0^{t_1} u_1(\tau) \sin \tau \, d\tau, \quad r \sin \alpha = -\int_0^{t_1} u_1(\tau) \cos \tau \, d\tau.$$

Deduce that, if $t_1 = \pi$ and α is kept fixed, all points of the line $-2 \leq r \leq 2$ are in $\mathscr{C}(\pi, \mathbf{0})$. Hence show that $\mathscr{C}(\pi, \mathbf{0})$ is the closed disc $\|\mathbf{x}\| \leq 2$.

Find $\mathscr{C}(n\pi, \mathbf{0})$ and show that $\mathscr{C}(\mathbf{0}) = \mathscr{R}^2$. (This exercise gives a direct proof of complete controllability for a system in which the eigenvalues of $\mathbf{A}$ have zero real parts. See §3.3(4).)

4 Time-optimal control

Dost thou love life? Then do not squander time.
Benjamin Franklin

This chapter contains the heart of Part A. We shall prove the existence of a time-optimal control and find how it can be determined. We shall consider the problem of the uniqueness of the optimal solution.

4.1. A balancing problem

A tightrope walker seeks to maintain an upright position; any deviation from the vertical will result in a catastrophic fall unless it is controlled, and the acrobat's objective is to regain the (unstable) equilibrium position as quickly as possible. The dynamics of this system is quite complicated and multi-dimensional. A simple model that demonstrates the main features of the control of an unstable equilibrium has a state equation of the form

$$\dot{x}_1 = x_1 + u_1. \tag{4.1}$$

With $u_1 = 0$, the solution $x_1 = 0$ is possible, but any initial non-zero value of x_1 leads to exponential growth of x_1. Suppose then that we choose an initial value $x_1^0 = c > 0$ and try to find a control $u_1(t)$ that will steer x_1 to 0 in the shortest possible time. We suppose that the control is bounded, so that $|u_1(t)| \leq 1$. The solution of (4.1) is easily found to be

$$x_1(t) = \mathrm{e}^t\left(c + \int_0^t \mathrm{e}^{-\tau} u_1(\tau)\,\mathrm{d}\tau\right). \tag{4.2}$$

From the result, it is clear that the set reachable from the initial point at time t_1 is the closed interval

$$(c - 1)\exp(t_1) + 1 \leq x_1^1 \leq (c + 1)\exp(t_1) - 1, \tag{4.3}$$

and that the end points of this interval are reached by the application of controls equal to -1 and $+1$, respectively, for $0 < t < t_1$. Provided $c < 1$, this set contains the origin when $t_1 \geq -\ln(1 - c)$ and does not contain the origin when $0 \leq t_1 < -\ln(1 - c)$. Hence equilibrium can be regained in the optimal time $-\ln(1 - c)$ by application of the control $u_1(t) = -1$.

Of course, if $c \geq 1$ the initial state is not controllable. Note that the reachable set is continuous in t_1 and that this property is crucial to the existence of an optimal solution. For suppose that a reachable set consisted of the following intervals: $0 \leq x_1 \leq t_1$ for $0 \leq t_1 \leq 1$; $0 \leq x_1 \leq 2t_1$ for $t_1 > 1$. Then $x_1 = 1.5$ belongs to the reachable set for all $t_1 > 1$, but is not reachable for $t_1 \leq 1$, and there is no optimal time.

This example exhibits many of the features of the general case, notably that the optimal time can be found by considering the least value of the time for which the target belongs to the reachable set, and that the optimal control takes only its extreme values.

4.2. Statement of the general time-optimal problem

As in §3.2, we consider the linear autonomous system (3.3). The initial point is $\mathbf{x}^0$ and the target is $\mathbf{x}^1 = \mathbf{x}(t_1)$. The cost function $J = \int_0^{t_1} 1 \, \mathrm{d}t = t_1$ and the control $\mathbf{u}$ is taken from the set $\mathscr{U}_\mathrm{u}$ or from $\mathscr{U}_\mathrm{b}$. If the target point is the origin, the controllable set $\mathscr{C}(t_1)$ is the set of points $\mathbf{x}^0$ defined by

$$\mathbf{x}^0 = -\int_0^{t_1} \exp(-\mathbf{A}\tau)\mathbf{B}\mathbf{u}(\tau) \, \mathrm{d}\tau. \tag{4.4}$$

With a given initial point $\mathbf{x}^0$ that belongs to $\mathscr{C}$ and so is controllable for some values of t_1, we have to find the optimal time t_*, that is the least value of t_1 for which $\mathbf{x}^0 \in \mathscr{C}(t_1)$. We also want to find the corresponding optimal control $\mathbf{u}^*$. Instead of working with the controllable set defined by (4.4), we could make use of the reachable set defined by (3.32).

We can deal with the case of unbounded controls immediately. There is no optimal solution for controls $\mathbf{u} \in \mathscr{U}_\mathrm{u}$. If rank $\mathbf{M} = n$, where $\mathbf{M}$ is the controllability matrix, it was shown in §3.3(2) that $\mathscr{C}(t_1) = \mathscr{R}^n$ for all $t_1 > 0$. But $\mathscr{C}(0)$ consists of the single point $\mathbf{0}$, so there is no optimal time for any point $\mathbf{x}^0 \in \mathscr{R}^n$ other than $\mathbf{0}$. If rank $\mathbf{M} = k < n$, $\mathscr{C}(t_1)$ lies in an unbounded k-dimensional subspace of $\mathscr{R}^n$

and the same argument for the non-existence of an optimal time can be used. Of course, all this means is that the real problem has not been formulated sensibly. An arbitrarily small time is all that is needed to reach the target when there is no limit to the size of the controls that can be employed. Because unbounded controls fail to provide a sensible problem, we can restrict attention to the set $\mathscr{U}_b$ of bounded integrable controls from now until the end of Chapter 5.

4.3. The time-optimal maximum principle (TOP)

We prove first that a time-optimal solution exists. Using the formulation of the problem in §4.2, we define

$$t_* = \inf\{t_1 \geq 0 \,|\, \mathbf{x}^0 \in \mathscr{C}(t_1)\}. \tag{4.5}$$

Since $\mathbf{x}^0 \in \mathscr{C}$, the set of values of t_1 is non-empty, and it is bounded below by zero, so the infimum exists. From the definition of t_* it is clear that $\mathbf{x}^0 \in \mathscr{C}(t_1)$ for all $t_1 > t_*$ and that $\mathbf{x}^0 \notin \mathscr{C}(t_1)$ for all $t_1 < t_*$. If we can show that $\mathbf{x}^0 \in \mathscr{C}(t_*)$, we shall have proved the existence of the optimal solution. Suppose that $\mathbf{x}^0 \notin \mathscr{C}(t_*)$. Since $\mathscr{C}(t_*)$ is closed, $\mathbf{x}^0$ can be separated from $\mathscr{C}(t_*)$ by a hyperplane and it is the centre of a ball $\mathscr{B}(\mathbf{x}^0, r)$ which has no intersection with $\mathscr{C}(t_*)$. Now increase the value of t_* by a small amount δ; because $\mathscr{C}(t)$ is continuous, we can shrink the ball to a radius r_1, if necessary, and retain an empty intersection between $\mathscr{C}(t_* + \delta)$ and $\mathscr{B}(\mathbf{x}^0, r_1)$. Hence $\mathbf{x}^0 \notin \mathscr{C}(t_* + \delta)$ which contradicts (4.5). Thus $\mathbf{x}^0 \in \mathscr{C}(t_*)$ and t_* is the optimal time. It also follows similarly that $\mathbf{x}^0 \in \partial\mathscr{C}(t_*)$.

This existence proof gives no indication of how the optimal time and the associated optimal control can be determined. To do this, it is necessary to consider all the points that are reachable from $\mathbf{x}^0$ over the whole time interval from 0 to t_*. Adding time to the n space dimensions gives us a set $\mathscr{RT}(t_*, \mathbf{x}^0)$ in $\mathscr{R}^{n+1}$, consisting of the points in $\mathscr{R}(t, \mathbf{x}^0)$, $0 \leq t \leq t_*$. The cross-section of this set by any plane t = constant is the reachable set at that time. Any control for which the response, that is, the solution of the state equations using this control, lies on the boundary of the $\mathscr{RT}$ set is called an *extremal control*. In the illustration in Fig. 4.1 for $n = 1$, the extremal controls are those for which the state x_1 takes the paths from A to B and to C along the boundaries of the triangular $\mathscr{RT}$ set.

We now prove that the time-optimal control is extremal. We do this by showing that if the response at any time t_2 is in the interior of

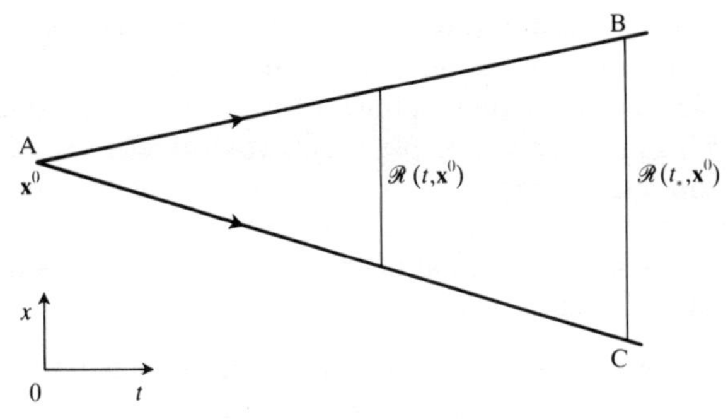

Fig. 4.1. The triangle ABC forms the $\mathscr{RT}$ set.

$\mathscr{R}(t_2, \mathbf{x}^0)$, then it remains in the interior of $\mathscr{R}(t, \mathbf{x}^0)$ for all later times $t > t_2$. Since $\mathbf{0} \in \partial\mathscr{R}(t_*, \mathbf{x}^0)$, the response using the optimal control must lie on $\partial\mathscr{RT}$ and the time-optimal control is extremal. Let $\mathbf{x}(t)$ be the response at time t, using the optimal control $\mathbf{u}^*$, and suppose that $\mathbf{x}(t_2) \in \text{Int}\, \mathscr{R}(t_2, \mathbf{x}^0)$. We can surround $\mathbf{x}(t_2)$ by an open ball $\mathscr{B}(\mathbf{x}(t_2), r_2)$ lying in $\mathscr{R}(t_2, \mathbf{x}^0)$. Apply the optimal control to all the points in this ball. At any time $t_3 > t_2$ they will lie in an open ball $\mathscr{B}(\mathbf{x}(t_3), r_3)$ and they must also belong to $\mathscr{R}(t_3, \mathbf{x}^0)$ since they have been reached by use of an admissible control. Thus, they belong to Int $\mathscr{R}(t_3, \mathbf{x}^0)$. See Fig. 4.2 for $n = 2$, in which the ball has become a disc.

We have therefore proved that if the response, using the optimal control, were to leave the boundary of $\mathscr{RT}$, it would remain in the interior for all later times and could not possibly reach the target

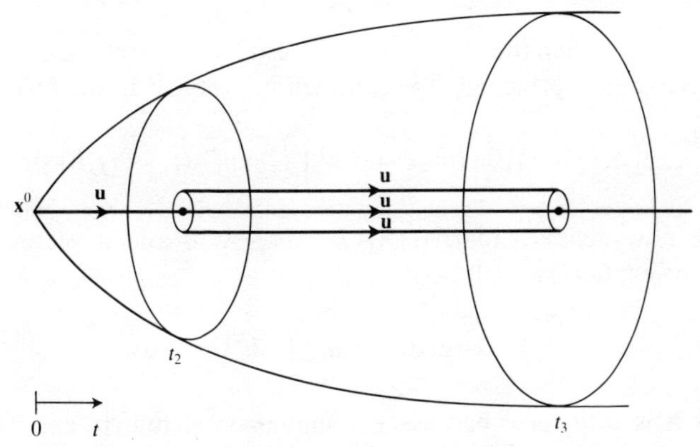

Fig. 4.2. Trajectories in the interior of $\mathscr{RT}$ remain in the interior.

point, which is known to lie on the boundary. The time-optimal control is therefore an extremal one. We now show how the extremal controls, including the time-optimal control, can be identified from the following maximum principle, which we shall refer to as TOP, for the time-optimal principle.

The control $\mathbf{u}(t)$ is extremal if and only if there exists a non-zero vector $\mathbf{h}$ such that, for $0 \leq t \leq t_1$,

$$\mathbf{h}^{\mathrm{T}}\exp(-\mathbf{A}t)\mathbf{B}\mathbf{u}(t) = \sup_{\mathbf{v}(t)\in\mathscr{U}_{\mathrm{b}}} \mathbf{h}^{\mathrm{T}}\exp(-\mathbf{A}t)\mathbf{B}\mathbf{v}(t). \tag{4.6}$$

Because the control set is $\mathscr{U}_{\mathrm{b}}$, the components of the extremal control are then given by

$$u_i(t) = \operatorname{sgn}[\mathbf{h}^{\mathrm{T}} \exp(-\mathbf{A}t)\mathbf{B}]_i, \quad \text{for } i = 1, 2, \ldots, m. \tag{4.7}$$

Suppose first that $\mathbf{u}$ is extremal and let $\mathbf{x}(t)$ be the corresponding solution of the state equations which reaches the target $\mathbf{x}^1$ at time t_1, so that $\mathbf{x}(t_1) = \mathbf{x}^1 \in \partial\mathscr{R}(t_1, \mathbf{x}^0)$. The reachable set is convex, so, from §2.6, there is a supporting hyperplane with normal $\mathbf{n}$ such that

$$\mathbf{n}\,.\,\mathbf{x}^1 = \sup_{\mathbf{y}^1\in\mathscr{R}} \mathbf{n}\,.\,\mathbf{y}^1, \tag{4.8}$$

where $\mathscr{R}$ is an abbreviation for $\mathscr{R}(t_1, \mathbf{x}^0)$. If $\mathbf{v}$ is an arbitrary control, and $\mathbf{y}$ the corresponding solution of the state equations, we have, from (3.32),

$$\mathbf{y}^1 = \exp(\mathbf{A}t_1)\left(\mathbf{x}^0 + \int_0^{t_1} \exp(-\mathbf{A}\tau)\mathbf{B}\mathbf{v}\,\mathrm{d}\tau\right), \tag{4.9}$$

and $\mathbf{y}^1 = \mathbf{x}^1$ when the control $\mathbf{v}$ is replaced by the extremal control $\mathbf{u}$. The maximum principle (4.8) can then be restated in the form

$$\mathbf{n}^{\mathrm{T}}\int_0^{t_1} \exp[\mathbf{A}(t_1 - \tau)]\mathbf{B}\mathbf{u}\,\mathrm{d}\tau = \sup_{\mathbf{v}\in\mathscr{U}_{\mathrm{b}}} \mathbf{n}^{\mathrm{T}}\int_0^{t_1} \exp[\mathbf{A}(t_1 - \tau)]\mathbf{B}\mathbf{v}\,\mathrm{d}\tau. \tag{4.10}$$

If we now define a matrix $\mathbf{Q}(t) = \exp(-\mathbf{A}t)\mathbf{B}$ and a vector $\mathbf{h}$ by $\mathbf{h}^{\mathrm{T}} = \mathbf{n}^{\mathrm{T}} \exp(\mathbf{A}t_1)$, (4.10) becomes

$$\int_0^{t_1} \mathbf{h}^{\mathrm{T}}\mathbf{Q}(\tau)\mathbf{u}\,\mathrm{d}\tau = \sup_{\mathbf{v}\in\mathscr{U}_{\mathrm{b}}} \int_0^{t_1} \mathbf{h}^{\mathrm{T}}\mathbf{Q}(\tau)\mathbf{v}\,\mathrm{d}\tau. \tag{4.11}$$

Now $\mathbf{h}$ is non-zero, because the fundamental matrix $\exp(\mathbf{A}t_1)$ is non-singular, and to achieve the supremum in (4.11) we must maximize the integrand at all times t between 0 and t_1. Moreover, since

$$\int_0^{t_1} \mathbf{h}^{\mathrm{T}}\mathbf{Q}(\tau)\mathbf{v}\,\mathrm{d}\tau = \sum_1^m \int_0^{t_1} [\mathbf{h}^{\mathrm{T}}\mathbf{Q}(\tau)]_i v_i\,\mathrm{d}\tau, \tag{4.12}$$

the supremum is achieved by taking

$$u_i = +1 \quad \text{when } [\mathbf{h}^{\mathrm{T}}\mathbf{Q}(\tau)]_i > 0, \quad u_i = -1 \quad \text{when } [\mathbf{h}^{\mathrm{T}}\mathbf{Q}]_i < 0, \tag{4.13}$$

or, in other words,

$$u_i(t) = \operatorname{sgn}[\mathbf{h}^{\mathrm{T}}\mathbf{Q}(t)]_i \quad \text{for } 0 \le t \le t_1. \tag{4.14}$$

We have therefore established the condition that the extremal control must satisfy. To prove the converse, the steps of the argument can be retraced in reverse order. We have now proved that an optimal solution exists when we are working with controls belonging to $\mathscr{U}_{\mathrm{b}}$, and we have proved by (4.14) that the optimal control must be bang–bang. It follows that it is unnecessary to consider separately the control set $\mathscr{U}_{\mathrm{bb}}$, since this is a subset of $\mathscr{U}_{\mathrm{b}}$ and the optimal control belongs to $\mathscr{U}_{\mathrm{bb}}$.

It is not at first obvious that this maximum principle TOP is enough to enable us to find the optimal solution. The vector $\mathbf{h}$ is unknown, and so is the time t_1. Of course, if we could determine the reachable set explicitly there would be no difficulty. The normal to the supporting hyperplane would be known and it would be easy to determine the time at which the target point first belonged to the reachable set. But it is not always (indeed rarely) possible to find the reachable set explicitly. As well as the maximum principle, though, we have the additional information that the optimal control must enable us to reach the target; these two facts, as we shall see in the examples, are sufficient to enable us to determine the solution.

Before we go on to consider some simple applications of these results, we should note that the component u_i of an extremal control is not determined by (4.14) when $[\mathbf{h}^{\mathrm{T}}\mathbf{Q}(t)]_i = p(t)$, say, is zero. Since $\mathbf{Q}(t)$ is an analytic function of t, there are only two possibilities. In the first case, $p(t)$ has a finite number of zeros in the interval from 0 to t_1. The control u_i is determined uniquely except at these discrete times, that is, almost everywhere, and the optimal control is bang–bang and has a finite number of switches so that it is piecewise constant. The values assigned to u_i at these switching times do not affect the trajectory; if desired, we could set them equal to $+1$ and obtain an optimal control that is bang–bang everywhere. In the second case, $p(t) = 0$ for all t between 0 and t_1, and then u_i is arbitrary. We call problems of the first kind *normal* and those in which the second possibility can occur *singular*. We shall discuss normality later, in §4.5, and see how it is related to the question of the uniqueness of the optimal control.

Another point that needs attention is that the maximum principle was derived for extremal controls. The optimal solution is included in this category, but it is possible for other controls to be extremal; the time taken to reach the target using such controls may not be the smallest possible and may, for example, give us a *maximum* time. Such possibilities arise when the conditions that determine $\mathbf{h}$ and t_1 have more than one solution. All solutions must be examined to determine which of them gives the least value of t_1, which can then be identified as the optimal time t_*. An example of this type of problem is included in the following section.

4.4. Simple examples of time-optimal problems

We can now demonstrate how the results of the previous section can be used to determine the solution of time-optimal problems in some simple cases.

1. *The balancing problem*

As a very simple introductory example, let us see how the general theory applies to the problem posed in §4.1, for which it was easy to construct the reachable set. In this one-dimensional example, $\exp(-\mathbf{A}t) = \mathrm{e}^{-t}$ and $\mathbf{B} = 1$. If $\mathbf{h} = \alpha$, the condition (4.14) satisfied by extremal controls becomes

$$u_1 = \mathrm{sgn}(\alpha \mathrm{e}^{-t}). \tag{4.15}$$

There are just two possibilities: either $\alpha > 0$ and $u_1 = +1$ for all t, or $\alpha < 0$ and $u_1 = -1$ for all t. We know from the maximum principle that $\mathbf{h}$ is a non-zero vector, so the possibility that $\alpha = 0$ does not arise. If $u_1 = +1$ throughout, the solution of the state equation (4.1), with $x_1(0) = x_1^0 = c$ and $0 < c < 1$, is

$$x_1(t) = (c + 1)\mathrm{e}^t - 1 \tag{4.16}$$

and the target $x_1^1 = 0$ is never reached. This alternative can therefore be dismissed; it provides an extremal control, but not the required optimal solution since the target is not hit. If $u_1 = -1$ throughout,

$$x_1(t) = (c - 1)e^t + 1 \tag{4.17}$$

and the target is reached at $t_1 = -\ln(1 - c)$. Since this is the only solution that satisfies (4.14) and enables the target to be reached, and we know that an optimal solution exists, we can be sure that the optimal time $t_* = -\ln(1 - c)$, the optimal control $u_1^* = -1$, and the optimal trajectory is given by (4.17). The optimal control is bang–bang, but without any switches from one extreme value to the other.

Suppose we repeat this problem, but with an initial value $x_1^0 = 2$ and with the target point $x_1^1 = 3$. The extremal controls are $u_1 = +1$ and $u_1 = -1$, as before, and the corresponding extremal trajectories are

$$x_1(t) = 3e^t - 1 \quad \text{when } u_1 = +1, \quad x_1(t) = e^t + 1 \quad \text{when } u_1 = -1. \tag{4.18}$$

Both these trajectories reach the target; the first at time $t_1 = \ln \frac{4}{3}$ and the second at $t_1 = \ln 2$. We therefore have two acceptable candidates for the optimal solution and we must pick the one with the smaller value of t_1. Thus $t_* = \ln \frac{4}{3}$, $u_1^*(t) = +1$, and the optimal trajectory is the first equation in (4.18). Figure 4.3 shows the two extremal trajectories lying along the boundaries of the $\mathscr{RT}$ set.

These solutions are, of course, intuitively obvious. To increase x_1 so as to reach the target as quickly as possible, we apply a positive control all the time, and at maximum strength. The other extremal

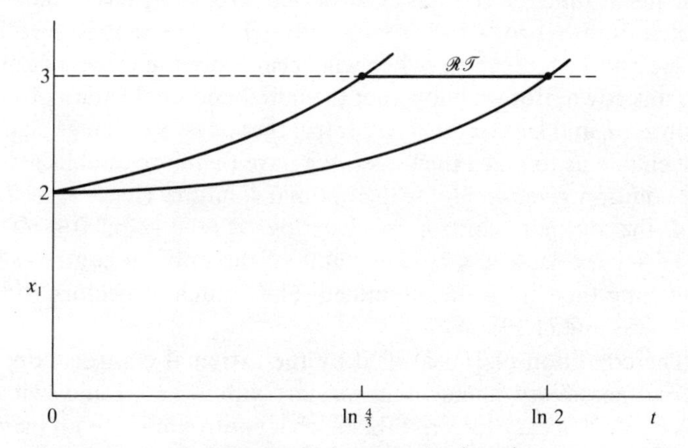

Fig. 4.3. Two extremal trajectories.

solution delays reaching the target for as long as possible, and $t_1 = \ln 2$ is the solution of the maximum-time problem.

2. *The positioning problem*

In §1.4 we defined the positioning problem and saw that it was possible to reach the target using bang–bang controls with a single switch. It was not possible to prove there that the solution obtained was the time-optimal one. With the principle established in §4.3 we can reconsider this problem and determine the optimal solution. The positioning problem is a two-dimensional one, with a single control, so $n = 2$ and $m = 1$. We choose an initial point $\mathbf{x}^0 = (x_1^0, x_2^0)^{\mathrm{T}} = (X, 0)^{\mathrm{T}}$ with $X > 0$ and the target is $\mathbf{x}^1 = \mathbf{0}$. The state equations (1.13) are in the general form (3.3) for a linear autonomous problem, with

$$\mathbf{A} = \begin{bmatrix} 0 & 1 \\ 0 & 0 \end{bmatrix}, \quad \mathbf{B} = \begin{bmatrix} 0 \\ 1 \end{bmatrix}. \tag{4.19}$$

Since $\mathbf{A}^2 = \mathbf{0}$, all higher powers of $\mathbf{A}$ are also equal to $\mathbf{0}$ and

$$\exp(-\mathbf{A}t) = \mathbf{I} - \mathbf{A}t = \begin{bmatrix} 1 & -t \\ 0 & 1 \end{bmatrix}. \tag{4.20}$$

If we suppose that $\mathbf{h}^{\mathrm{T}} = (\alpha, \beta)$, where α and β are not both zero, the maximum principle (4.14) states that

$$u_1(t) = \operatorname{sgn}(\beta - \alpha t). \tag{4.21}$$

The linear function of t has at most one zero, so u_1 has at most one switch from $+1$ to -1 or from -1 to $+1$. We cannot be sure from (4.21) if and when such a switch will occur, since the values of α and β are unknown. But we know that the initial and final values of x_2 are both zero, and since $\dot{x}_2 = u_1$ a control that does not change sign will not enable us to reach the target. We have therefore established that the solution given in §1.4 is the optimal solution. Hence $t_* = 2X^{1/2}$ and the optimal control is given by $u_1^* = -1$ for $0 \le t < \frac{1}{2}t_*$, $u_1^* = +1$ for $\frac{1}{2}t_* < t \le t_*$. The value of the optimal control at the switching time $\frac{1}{2}t_*$ is undetermined. The optimal trajectory is given by (1.18) and (1.19).

The condition (4.21) satisfied by the extremal controls admits a second possibility, namely that we start with $u_1 = +1$ and switch to $u_1 = -1$. It is easy to verify that such a control leads to an increase in the value of x_1, and when $x_2 = 0$ we are further from the target than when we started. This sequence of controls does not enable the target to be reached, and so can be dismissed. We have, therefore, found a unique optimal control that will lead to the target in the shortest possible time.

These elementary examples have demonstrated that the maximum principle provides information about the number of switches the optimal control, which is known to be bang–bang, can take. This information, combined with the solutions of the state equations using such bang–bang controls, enables us to identify all those extremal solutions for which the target point can be successfully attained. If there is more than one such solution, then we choose the one that reaches the target in the shortest time. It is possible, of course, that this procedure does not identify the optimal control uniquely, although the optimal time will be determined. Some more complicated examples are discussed in the next chapter, where the strategy we have developed is found to be sufficient to enable their solutions to be found. Now we take up the question of the uniqueness of the optimal control, as promised in §4.3.

4.5. Uniqueness and normality

There is a trivial non-uniqueness in the optimal control that results from the failure of the maximum principle

$$u_i = \text{sgn}[\mathbf{h}^{\text{T}}\exp(-\mathbf{A}t)\mathbf{B}]_i \tag{4.22}$$

to determine u_i at the instants when it switches between the extreme values -1 and $+1$. Values at these instants may be assigned arbitrarily without affecting the optimal time or the optimal trajectory. What is significant is the lack of information provided by (4.22) when the argument of the sgn function is zero over an interval. Because of the analyticity of $\exp(-\mathbf{A}t)$, it must then vanish for all t and the corresponding component of the control vector is unrestricted. As an example of what can happen in that case, consider the system of Exercise 3.5.

The state has two components which remain constant unless acted on by controls which affect each component independently. The state equation is $\dot{\mathbf{x}} = \mathbf{u}$ and both the state and control vectors are two-dimensional, so $m = n = 2$. Then $\mathbf{A} = \mathbf{0}$ and $\mathbf{B} = \mathbf{I}$ and, if $\mathbf{h}^{\text{T}} = (\alpha, \beta)$, we see from (4.22) that

$$u_1 = \text{sgn}(\alpha), \quad u_2 = \text{sgn}(\beta). \tag{4.23}$$

If neither α nor β is zero, the control is uniquely determined. The various possible signs give four possibilities. With $\mathbf{x}^0 = \mathbf{0}$, the optimal trajectories will be given by

$$x_1(t) = \int_0^t (\pm 1)\, \mathrm{d}\tau = \pm t, \quad x_2(t) = \pm t, \tag{4.24}$$

and the controls to reach the four target points $(\pm 1, \pm 1)$ in the optimal time $t_* = 1$ are unique. If, however, either α or β is zero, one of the controls is undetermined. Suppose, for example, the target point is $x_1^1 = 1$, $x_2^1 = c$, with $-1 < c < 1$. This point can be reached in the optimal time $t_* = 1$ with the control $u_1 = 1$ and with any control u_2 provided that

$$c = \int_0^1 u_2(t)\, \mathrm{d}t. \tag{4.25}$$

There are many possible choices for u_2, including bang–bang ones. This target point is associated with the choice $\alpha > 0$, $\beta = 0$, and the other possible choices correspond to target points on the other sides of the square with vertices $(\pm 1, \pm 1)$. This square, sketched in Fig. 4.4, forms the reachable set at time $t = 1$ for the initial point at the origin. It is not strictly convex and the only extreme points (see §2.6) are the four vertices. Because $\mathbf{A} = \mathbf{0}$, $\mathbf{h} = \mathbf{n}$, where $\mathbf{n}$ is the normal to the supporting line at any point on the boundary of the set. At the extreme point A, with coordinates (1, 1), this normal is not unique; $\mathbf{n}^{\mathrm{T}} = (\alpha, \beta)$ with α and β both positive and the optimal control *is* unique. At the non-extreme point B, with coordinates $(1, c)$, $\mathbf{n}^{\mathrm{T}} = (1, 0)$ and the optimal control is *not* uniquely determined, as we have already seen.

This example suggests a connection between the uniqueness of the optimal control and the target being an extreme point of the

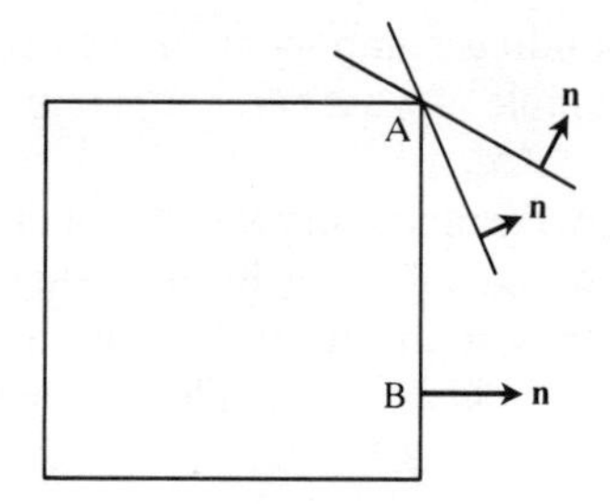

Fig. 4.4. The reachable set at time $t = 1$.

reachable set. Since it is only for strictly convex sets that all boundary points are extreme points, we may expect non-uniqueness of the optimal control to be associated with reachable sets that are not strictly convex.

To support these conjectures, and to find a criterion to determine whether or not a unique optimal control exists, we define a normal problem in terms of the matrices $\mathbf{A}$ and $\mathbf{B}$ of the linear system. Let $\mathbf{b}^{(i)}$ be the ith column of $\mathbf{B}$, $i = 1, 2, \ldots, m$, and define the $n \times n$ matrices $\mathbf{M}^{(i)}$ by

$$\mathbf{M}^{(i)} = [\mathbf{b}^{(i)} \quad \mathbf{A}\mathbf{b}^{(i)} \quad \mathbf{A}^2\mathbf{b}^{(i)} \quad \ldots \quad \mathbf{A}^{n-1}\mathbf{b}^{(i)}]. \tag{4.26}$$

The columns of this collection of matrices can be rearranged to form the controllability matrix $\mathbf{M}$ (3.14), and we saw in §3.3 that the rank of $\mathbf{M}$ was associated with the controllability of the system. Normality is associated with the ranks of each of the matrices $\mathbf{M}^{(i)}$ and we say that the problem is *normal* if and only if

$$\text{rank } \mathbf{M}^{(i)} = n \quad \text{for } i = 1, 2, \ldots, m. \tag{4.27}$$

Clearly, if any one of these matrices has rank equal to n, $\mathbf{M}$ contains n independent columns and also has rank equal to n. Thus the condition for normality is more stringent than that for controllability.

In exactly the same way as in the proof of controllability in §3.3, we can show that rank $\mathbf{M}^{(i)} = n$ if and only if there is no non-zero vector $\mathbf{h}$ such that $\mathbf{h}^{\mathrm{T}} \exp(-\mathbf{A}\tau)\mathbf{b}^{(i)} = 0$ for $0 \leq \tau \leq t_1$. It follows at once that (4.22) determines u_i uniquely for a normal system. In the example that we have just considered,

$$\mathbf{M}^{(1)} = \begin{bmatrix} 1 & 0 \\ 0 & 0 \end{bmatrix}, \quad \mathbf{M}^{(2)} = \begin{bmatrix} 0 & 0 \\ 1 & 0 \end{bmatrix}, \tag{4.28}$$

both of which have rank 1, so the system is not normal, although

$$\mathbf{M} = \begin{bmatrix} 1 & 0 & 0 & 0 \\ 0 & 1 & 0 & 0 \end{bmatrix}, \tag{4.29}$$

which has rank 2. Notice also from this example that a unique optimal control does not imply normality. The optimal control to the point (1,1) is unique, but the system is not normal. As we shall see, it is only when there is a unique optimal control to each point of the boundary of the reachable set that we can infer that the system is normal. If the reachable set has collapsed on to a hyperplane of dimension less than n in $\mathscr{R}^n$, we shall here define ∂ and Int relative to this hyperplane. Suppose, for example, that the reachable set were the closed disc $\mathscr{D}$ in $\mathscr{R}^3$, defined by $x_1^2 + x_2^2 \leq 1$, $x_3 = 0$; relative to the plane $x_3 = 0$ in which the set lies, $\partial\mathscr{D}$ is the circle $x_1^2 + x_2^2 = 1$ and Int $\mathscr{D}$ is the open disc $x_1^2 + x_2^2 < 1$.

We can abbreviate the notation $\mathscr{R}(t_1, \mathbf{x}^0)$ for the reachable set at time t_1 from the initial point $\mathbf{x}^0$ to $\mathscr{R}$. We first show that, if $\mathscr{R}$ is not strictly convex, then the control leading to a non-extreme point of $\partial\mathscr{R}$ is not unique. For if $\mathbf{x}$ is not an extreme point of $\partial\mathscr{R}$, by definition there are two other points of $\partial\mathscr{R}$, say $\mathbf{y}$ and $\mathbf{z}$ such that $\mathbf{x} = \frac{1}{2}(\mathbf{y} + \mathbf{z})$. From the bang–bang principle (and this is the first occasion on which we have made use of it) there are bang–bang controls $\mathbf{u}$, $\mathbf{v}$, and $\mathbf{w}$, say, leading respectively to $\mathbf{x}$, $\mathbf{y}$, and $\mathbf{z}$. From the linearity of the reachable set, it follows that the control $\hat{\mathbf{u}} = \frac{1}{2}(\mathbf{v} + \mathbf{w})$ will also lead to the target point $\mathbf{x}$. Now $\hat{\mathbf{u}}$ cannot be a bang–bang control, for $\mathbf{v}$ and $\mathbf{w}$ must differ in at least one component over some interval of t (or over some set of non-zero measure), and if $v_i = +1$ and $w_i = -1$, then $\hat{u}_i = 0$ and similarly if the signs are reversed. Therefore $\hat{\mathbf{u}}$ is not bang–bang and we have two different controls, $\mathbf{u}$ and $\hat{\mathbf{u}}$, both leading to the same target point $\mathbf{x}$.

Now suppose that there are two controls $\mathbf{u}$ and $\mathbf{v}$, both leading to a point $\mathbf{x}^1$ of $\partial\mathscr{R}$, and that the trajectories corresponding to these controls are $\mathbf{x}(t)$ and $\mathbf{y}(t)$, with $\mathbf{x}(t_1) = \mathbf{y}(t_1) = \mathbf{x}^1$. We also suppose that these trajectories are different, and we shall show that $\mathbf{x}^1$ cannot be an extreme point of $\mathscr{R}$, and therefore $\mathscr{R}$ cannot be strictly convex. Let $\mathbf{x}^2$ and $\mathbf{y}^2$ be points on these two trajectories at $t = t_2 < t_1$, and $\mathbf{x}^2 \neq \mathbf{y}^2$ (note that the superscript 2 does not denote the power). For $t_2 < t \leq t_1$, switch the controls round and apply the control $\mathbf{v}$ for a trajectory starting at $\mathbf{x}^2$ and the control $\mathbf{u}$ for a trajectory starting at $\mathbf{y}^2$ (see Fig. 4.5). At time t_1 these trajectories will reach the points $\mathbf{p}^1$ and $\mathbf{q}^1$, say, which belong to $\mathscr{R}$ since the controls used are admissible.

These two points can be found from the general solution of the state equations (3.32) and are given by the following expressions:

$$\mathbf{p}^1 = \exp(\mathbf{A}t_1)\left(\exp(-\mathbf{A}t_2)\mathbf{x}^2 + \int_{t_2}^{t_1} \exp(-\mathbf{A}\tau)\mathbf{B}\mathbf{v}\,\mathrm{d}\tau\right), \quad (4.30)$$

$$\mathbf{q}^1 = \exp(\mathbf{A}t_1)\left(\exp(-\mathbf{A}t_2)\mathbf{y}^2 + \int_{t_2}^{t_1} \exp(-\mathbf{A}\tau)\mathbf{B}\mathbf{u}\,\mathrm{d}\tau\right). \quad (4.31)$$

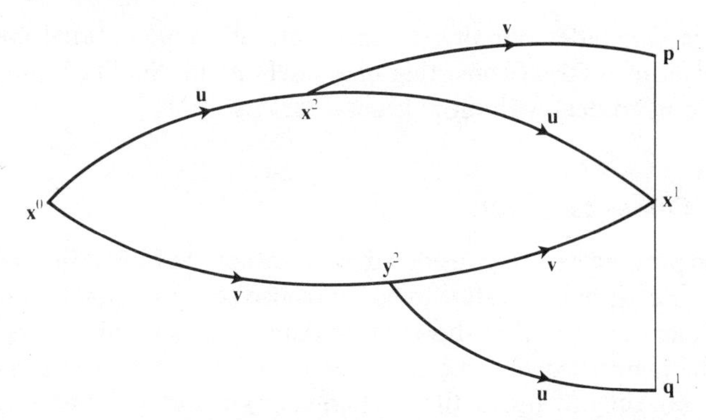

Fig. 4.5. Two controls leading to a boundary point that is not an extreme point.

If we use the original controls, both trajectories lead to the same point $\mathbf{x}^1$, of course, and we have the corresponding results:

$$\mathbf{x}^1 = \exp(\mathbf{A}t_1)\left(\exp(-\mathbf{A}t_2)\mathbf{x}^2 + \int_{t_2}^{t_1} \exp(-\mathbf{A}\tau)\mathbf{B}\mathbf{u}\,\mathrm{d}\tau\right), \quad (4.32)$$

$$\mathbf{x}^1 = \exp(\mathbf{A}t_1)\left(\exp(-\mathbf{A}t_2)\mathbf{y}^2 + \int_{t_2}^{t_1} \exp(-\mathbf{A}\tau)\mathbf{B}\mathbf{v}\,\mathrm{d}\tau\right). \quad (4.33)$$

It follows from these expressions that

$$\mathbf{x}^1 - \mathbf{p}^1 = \mathbf{q}^1 - \mathbf{x}^1 \neq \mathbf{0} \quad (4.34)$$

and hence

$$\mathbf{x}^1 = \tfrac{1}{2}(\mathbf{p}^1 + \mathbf{q}^1). \quad (4.35)$$

Thus $\mathbf{x}^1$ is not an extreme point of $\mathscr{R}$ and so $\mathscr{R}$ is not strictly convex. In fact, the three points $\mathbf{x}^1$, $\mathbf{p}^1$, and $\mathbf{q}^1$ all lie in $\partial\mathscr{R}$.

In proving this result, we have assumed that different controls cannot lead to the same response. This is certainly true if we have removed any redundancy from the controls, as explained in §3.5, although the result can be proved without this restriction. Since redundant controls can be removed without losing anything of significance in applications of the theory, we shall assume that this has been done. The interconnections between the normality of the system, the uniqueness of the optimal control, and the strict convexity of the reachable set have now been identified.

This concludes our investigation into the time-optimal control problem to a target consisting of a single point. Similar arguments enable us to deal with more general targets.

4.6. Convex target sets

In the problems so far considered, we have sought to find the optimal time for a given initial state to be controlled to a given final state. It is possible, however, that this is more than what is required. We may not be concerned to force the system to a completely determined state, but only to ensure that certain components of the state vector achieve desired values. Can we extend the methods we have developed in this chapter to cover such less constricted problems? First let us look at the following simple case.

1. *The braking problem*

In the positioning problem described in §1.4 and solved in §4.4(2) the target was the origin; both the position and the velocity of the object being moved had to achieve given values. However, in braking, the requirement is to reduce the velocity to zero, without necessarily needing to prescribe the final position. Suppose, using the notation of §4.4(2), that the initial state is $\mathbf{x}^0 = (0, Y)^{\mathrm{T}}$, with $Y > 0$, and that the objective is to force x_2 to zero in the shortest possible time. The target is, therefore, the line $x_2 = 0$ and the final value of x_1 is arbitrary. As before, we consider the set of points reachable from $\mathbf{x}^0$ in time t_1 as t_1 increases, using controls belonging to $\mathscr{U}_{\mathrm{b}}$. The solution of this problem is intuitively obvious. In order to reduce the speed to zero as quickly as possible, we apply the control $u_1 = -1$, and the associated response is given by $x_2 = Y - t$. The optimal time is therefore $t_* = Y$ and the target reached is the point $(\frac{1}{2}Y^2, 0)$.

We could have defined the target as the half-space $x_2 \leq 0$ and the first contact with this target set is at its boundary. When the target is a single point or, as in this example, a line or half-space, the optimal time is the smallest value of t for which the reachable set and the target intersect. This forms the basis for our extension of the maximum principle for general targets.

We suppose that the target $\mathscr{T}$ is some closed convex set in $\mathscr{R}^n$ and that the initial point $\mathbf{x}^0 \notin \mathscr{T}$. If $\mathscr{R}(t_1, \mathbf{x}^0)$, abbreviated to $\mathscr{R}$, is the set reachable from $\mathbf{x}^0$ in time t_1 with controls $\mathbf{u} \in \mathscr{U}_{\mathrm{b}}$, we suppose that there are points in $\mathscr{T}$ that are reachable in some finite time. We can then define

$$t_* = \inf[t_1 \geq 0 | \mathscr{R} \bigcap \mathscr{T} \neq \varnothing], \tag{4.36}$$

that is, the smallest value of t_1 for which the intersection between the reachable set and the target set is not empty. The set of times is not empty and is bounded below, so the infimum exists. Previously we identified the optimal time as the smallest value of t_1 for which the target point belonged to the reachable set. We can do the same here and identify the optimal time as the smallest value of t_1 for which the reachable set intersects the target set. As before, there is a supporting hyperplane to $\mathscr{R}$ at the optimal time, with normal **n**, and the maximum principle (4.8) still holds. The difference is that the target point $\mathbf{x}^1$ is now not known, but is any point belonging to $\mathscr{R} \cap \mathscr{T}$ at the optimal time. This loss of information is balanced by the additional fact that, since $\mathscr{T}$ is convex, the hyperplane with normal **n** must also be the supporting hyperplane for $\mathscr{T}$ with normal $-\mathbf{n}$. Examples of some of the cases that can occur are sketched in Fig. 4.6.

When the target set is bounded by a hyperplane of dimension $n-1$ or less, we are given some information about **n**, without knowing the exact position of the final state within the target. The maximum principle again has the form given by (4.7), and the application of the principle, together with the requirement that the final state belongs to the target set, enables the solution to be found.

As an example of the application of these general results, let us consider the simple braking problem again. With $Y > 0$, the reachable set for small values of t lies above the target set $x_2 \leq 0$, so the normal pointing out of $\mathscr{R}$ and into $\mathscr{T}$ is $\mathbf{n}^{\mathrm{T}} = (0, -1)$. The maximum principle for this problem shows that

$$u_1 = \operatorname{sgn}\{\mathbf{n}^{\mathrm{T}} \exp[\mathbf{A}(t_1 - t)]\mathbf{B}\} = \operatorname{sgn}\{[0 \quad -1][t_1 - t \quad 1]^{\mathrm{T}}\}, \tag{4.37}$$

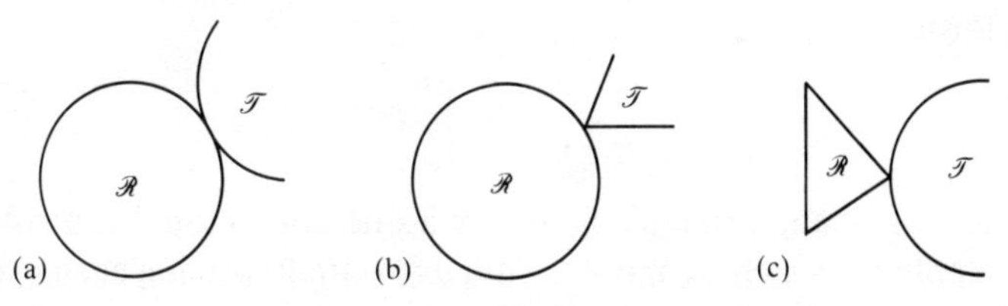

Fig. 4.6. Reachable and target sets.

where we have made use of (4.19) and (4.20). Hence the optimal control is $u_1^* = -1$, and, since $\dot{x}_2 = u_1$, the optimal trajectory is given by $x_2^* = Y - t$, $x_1^* = Yt - \frac{1}{2}t^2$, and the optimal time $t_* = Y$.

To conclude this chapter, and before going on to some applications of the theory we have derived, there is a topic postponed from Chapter 3 that we must consider. This is the bang–bang principle, defined in §3.4. We have used this principle in the proof of non-uniqueness for singular problems in §4.5 and we also need it to be certain that this non-uniqueness includes a bang–bang control. The existence of a bang–bang optimal solution for normal problems does not depend on this principle.

4.7. The bang–bang principle

The essence of the bang–bang principle for linear autonomous problems can be stated as follows. Suppose that $\mathbf{Q}(t)$ is an $n \times m$ matrix, with elements that are integrable functions of t for $0 \le t \le 1$ (we lose nothing by setting $t_1 = 1$). Let $\mathbf{x}$ be an arbitrary n-dimensional vector for which there exists an m-dimensional vector $\mathbf{v}(t)$, with integrable components satisfying $|v_i(t)| \le 1$ for $0 \le t \le 1$, such that

$$\mathbf{x} = \int_0^1 \mathbf{Q}(t)\mathbf{v}(t)\,\mathrm{d}t. \tag{4.38}$$

Then there is an m-dimensional vector $\mathbf{u}(t)$, with integrable components satisfying $|u_i(t)| = 1$ for $0 \le t \le 1$ such that

$$\mathbf{x} = \int_0^1 \mathbf{Q}(t)\mathbf{u}(t)\,\mathrm{d}t. \tag{4.39}$$

Put another way, if $\mathscr{S}[\mathscr{U}_\mathrm{b}]$ is the set of values of the integral in (4.38) with $\mathbf{v} \in \mathscr{U}_\mathrm{b}$, and $\mathscr{S}[\mathscr{U}_\mathrm{bb}]$ the corresponding set when $\mathbf{v} \in \mathscr{U}_\mathrm{bb}$, then $\mathscr{S}[\mathscr{U}_\mathrm{b}] = \mathscr{S}[\mathscr{U}_\mathrm{bb}]$.

The first point to notice is that we can deal with the components of $\mathbf{v}$ one by one, for if $\mathbf{Q}^{(i)}$ is the ith column of $\mathbf{Q}$, we can write (4.38) in the form

$$\mathbf{x} = \sum_1^m \mathbf{x}^{(i)} = \sum_1^m \int_0^1 \mathbf{Q}^{(i)} v_i\, dt. \tag{4.40}$$

If the bang–bang principle holds for a one-dimensional control, we can apply it to each term in the sum in (4.40). Replacing each v_i by a control u_i satisfying $|u_i(t)| = 1$, we arrive at a control $\mathbf{u} \in \mathscr{U}_\mathrm{bb}$. The second point is that we need only consider normal problems. For when $m = 1$, and we have just seen that this is sufficient, the condition for the problem not to be normal is that

$$\mathbf{h}^{\mathrm{T}}\mathbf{Q}(t) = 0 \quad \text{for } 0 \leq t \leq 1, \tag{4.41}$$

where $\mathbf{h}$ is a non-zero vector. The possible values of $\mathbf{x}$ then lie in a hyperplane of dimension $k < n$, and when we apply the bang–bang principle to k independent components of $\mathbf{x}$, the remaining components will retain their correct values. Thus a non-normal problem of degree n is equivalent to a normal problem of lower degree.

It follows, therefore, that to establish the bang–bang principle for LA problems, it is sufficient to prove the following theorem. Let $\mathbf{x}$ be any n-dimensional vector for which an integrable function $v(t)$ exists, satisfying $|v(t)| \leq 1$ for $0 \leq t \leq 1$, such that

$$\mathbf{x} = \int_0^1 \mathbf{q}(t)v(t)\,\mathrm{d}t, \tag{4.42}$$

where the n components of $\mathbf{q}$ are linearly independent. Then a piecewise-constant function $u(t)$ exists, such that $|u(t)| = 1$ for $0 \leq t \leq 1$ and

$$\mathbf{x} = \int_0^1 \mathbf{q}(t)u(t)\,\mathrm{d}t. \tag{4.43}$$

The argument follows similar lines to those we used to establish the maximum principle for LA problems, except that instead of reducing the time until the target point is on the boundary of the accessible points, as we did before, we now restrict the control set until $\mathbf{x}$ is on the boundary. We start with the control set $\mathscr{U}_{\mathrm{b}}$; let $\mathscr{U}[s]$ be the set of bounded controls such that $u \in \mathscr{U}[s]$ if

$$u = -1 \quad \text{for } 0 \leq t \leq s, \quad u \in \mathscr{U}_{\mathrm{b}} \quad \text{for } s < t \leq 1. \tag{4.44}$$

Then $\mathscr{U}[0] = \mathscr{U}_{\mathrm{b}}$, and $\mathscr{U}[1]$ contains the single control $u = -1$ for $0 \leq t \leq 1$. Let $\mathscr{R}[s]$ be the set of values of the integral (4.42) for $v \in \mathscr{U}[s]$. It is clear that $\mathbf{x} \in \mathscr{R}[0]$ and that $\mathscr{R}[1]$ contains a single point (if this point coincides with $\mathbf{x}$, $u = -1$ is the required bang–bang control). Also, since $\mathscr{U}[s]$ is convex, so is $\mathscr{R}[s]$, and it is also compact and continuous in s. As s increases, $\mathscr{R}[s]$ decreases; in fact $\mathscr{R}[s_1] \subset \mathscr{R}[s_2]$ when $s_1 > s_2$, since $\mathscr{U}[s_1] \subset \mathscr{U}[s_2]$. We can therefore define s_* by

$$s_* = \sup_{0 \leq s \leq 1} \{s \,|\, \mathbf{x} \in \mathscr{R}[s]\}. \tag{4.45}$$

The set of values of s is not empty and is bounded above, so the supremum exists. Exactly as in the proof of the maximum principle for the time-optimal problem, $\mathbf{x} \in \partial\mathscr{R}[s_*]$ and there is a supporting hyperplane with normal $\mathbf{n}$ such that

$$\mathbf{n} \,.\, \mathbf{x} = \sup_{\mathbf{y} \in \mathscr{R}[s_*]} \mathbf{n} \,.\, \mathbf{y}. \tag{4.46}$$

It follows that, if $u \in \mathscr{U}[s_*]$ is a control for $\mathbf{x}$,

$$\int_{s_*}^{1} \mathbf{n}^{\mathrm{T}}\mathbf{q}(t)u(t)\,\mathrm{d}t = \sup_{v \in \mathscr{U}[s_*]} \int_{s_*}^{1} \mathbf{n}^{\mathrm{T}}\mathbf{q}(t)v(t)\,\mathrm{d}t, \tag{4.47}$$

and, because the controls are bounded, the supremum is achieved when

$$u(t) = \operatorname{sgn}[\mathbf{n}^{\mathrm{T}}\mathbf{q}(t)] \quad \text{for } s_* < t \leq 1. \tag{4.48}$$

Because the components of $\mathbf{q}$ are independent, this defines $u = \pm 1$ except at a finite number of points. Since we already know that $u = -1$ for $0 \leq t \leq s_*$, we have shown that a piecewise-constant bang–bang control exists for an arbitrary point $\mathbf{x}$.

This completes the development of the theory for time-optimal control problems, and in the next chapter we apply the time-optimal principle TOP to some problems that are more difficult than the illustrative examples so far encountered.

Exercises 4

(*Solutions on p. 222*)

4.1 Show that, when $m = 1$, a controllable problem is also normal.

Suppose we have a normal problem with $m = 1$ and that the eigenvalues of $\mathbf{A}$ are distinct and real. Show that the optimal control is given by the sign of an expression of the form

$$p_n(t) = \sum_{1}^{n} c_i \exp(-\lambda_i t),$$

where λ_i is an eigenvalue of $\mathbf{A}$ and c_i is a constant. Prove that the optimal control has at most $n - 1$ switches.

4.2 Fragile goods have to be transported a distance of 100 km. They are unable to withstand an acceleration or retardation greater than 10 ms^{-2}. Assuming that a straight path is possible and that there is no resistance to the motion, find the shortest possible time for delivery.

Find also the greatest speed attained on the journey.

4.3 The population of a rare species would disappear unless steps were taken to control it, but the present population is too large for its environment. The controls available are either to import specimens from zoos, or to allow culling of the wild animals. If the population at time t is $p(t)$, an equation that models the growth or decay of the population has the form

$$\dot{p} = -kp + f,$$

where k is positive and $-F \leq f \leq F$. Suppose that $p(0) = P$ and that the desired population is $\frac{1}{2}P$. What control strategy should be used in order to reduce the population to this level in the shortest possible time? Find the optimal time, and show that, if $P > 2F/k$, there are two extremal solutions, only one of which is time optimal.

Is it possible to hold the population at the desired level?

4.4 Consider the system defined in Exercise 4.3, but with the initial state P less than a desired level Q. Find the optimal strategy and the optimal time when $Q < F/k$. Comment on the consequences of setting $Q \geq F/k$.

4.5 Consider the two-dimensional problem introduced in §3.2:

$$\dot{x}_1 = x_1 + u_1, \quad \dot{x}_2 = x_2 + u_1,$$

with $\mathbf{x}^0 = (p, q)^T$ and $\mathbf{x}^1 = \mathbf{0}$. The control u_1 is bounded, so that $|u_1(t)| \leq 1$. Show that the maximum principle TOP provides conditions on the optimal control, but that a solution only exists when $p = q$ and when $-1 < p < 1$. (This example emphasizes that TOP alone does not guarantee that a time-optimal solution exists.)

4.6 A two-dimensional system with two bounded controls is defined by the state equations

$$\dot{x}_1 = u_1 + u_2, \quad \dot{x}_2 = u_2.$$

Show that the problem is not normal.

If $\mathbf{x}^0 = \mathbf{0}$, find bang–bang optimal controls and the optimal time when the target point is (a) $\mathbf{x}^1 = (2, 1)^T$, (b) $\mathbf{x}^1 = (2, 0)^T$. Are these controls unique? If not, find another optimal control.

4.7 Consider the positioning problem, §4.4(2), with $\mathbf{x}^0 = (X, Y)^T$ and $X > 0$. Suppose the target set is the half-space $x_1 \leq 0$. Find the optimal control and the optimal time. Find also where the optimal trajectory meets the target set.

5 Further examples

Please, sir, I want some more.

Charles Dickens

The theory developed in Chapter 4 was illustrated there by some simple examples. Now we can apply it to problems of a more complex nature.

5.1. Positioning with two controls

We have already considered the problem of positioning an object by means of a controlling force. Now let us suppose that as well as controlling the rate of change of the velocity of the object, we are also able to control its velocity directly. Thus the statement of the problem to be solved is the same as in §4.4(2), except that the state equations are now given by

$$\dot{x}_1 = x_2 + u_1, \quad \dot{x}_2 = u_2. \tag{5.1}$$

Both controls belong to $\mathscr{U}_b$, so that $|u_1| \leq 1$, $|u_2| \leq 1$, and the target is the origin. We want to determine the controls that will steer the initial point (X_1, X_2) to the origin in the shortest possible time.

It is easy to show that this problem is not normal. For if we write the state equations in their standard form, we find that the matrices **A** and **B** are given by

$$\mathbf{A} = \begin{bmatrix} 0 & 1 \\ 0 & 0 \end{bmatrix}, \quad \mathbf{B} = \begin{bmatrix} 1 & 0 \\ 0 & 1 \end{bmatrix}, \tag{5.2}$$

so that, applying the test for normality from §4.5,

$$\mathbf{M}^{(1)} = \begin{bmatrix} 1 & 0 \\ 0 & 0 \end{bmatrix}, \quad \mathbf{M}^{(2)} = \begin{bmatrix} 0 & 1 \\ 1 & 0 \end{bmatrix}, \tag{5.3}$$

and rank $\mathbf{M}^{(1)} = 1$, rank $\mathbf{M}^{(2)} = 2$. The ranks are not both equal to two, so the problem is not normal and the optimal controls, from some initial points, will not be unique. If $\mathbf{h}^{\mathrm{T}} = (\alpha, \beta)$, the maximum principle TOP of §4.3 shows that

$$u_1 = \operatorname{sgn}(\alpha), \quad u_2 = \operatorname{sgn}(\beta - \alpha t), \tag{5.4}$$

since $\exp(-\mathbf{A}t) = \mathbf{I} - \mathbf{A}t$ and $\mathbf{B} = \mathbf{I}$. It follows that u_2 is uniquely determined, because $\mathbf{h}$ is not a null vector, and so $\beta - \alpha t$ is not identically zero. However, if $\alpha = 0$, u_1 is not determined by TOP, while u_2 is then of constant sign.

Let us examine this last possibility and determine the initial states for which it can occur. Then if u_2 is equal to $+1$ for all t,

$$x_2 = X_2 + t, \tag{5.5}$$

and $x_2 = 0$ at the terminal time $t_1 = -X_2$. This is the optimal solution for negative values of X_2, provided it is possible to choose u_1 in such a way that x_1 can be steered from X_1 to 0 in the same time t_1. Positive values of X_2 require $u_2 = -1$ and can be treated in a similar manner. The solution of the first state equation in (5.1), using the value of x_2 from (5.5), can be written as

$$x_1 = X_1 + X_2 t + \tfrac{1}{2}t^2 + \int_0^t u_1(\tau)\,\mathrm{d}\tau, \tag{5.6}$$

and we can reach the target at time t_1 provided we can find an admissible control u_1 such that

$$\int_0^{-X_2} u_1\,\mathrm{d}\tau = \tfrac{1}{2}X_2^2 - X_1, \tag{5.7}$$

since $t_1 = -X_2$. Since $|u_1| \leq 1$, we see that the value of X_1 must be chosen so that it lies in the interval

$$\tfrac{1}{2}X_2^2 + X_2 \leq X_1 \leq \tfrac{1}{2}X_2^2 - X_2. \tag{5.8}$$

For values of X_1 satisfying these inequalities the optimal time is $-X_2$ and the optimal control $u_2 = 1$; the value of $u_1(t)$ must be chosen to satisfy (5.7) and the bounds $|u_1| \leq 1$. Except when X_1 is at the ends of the interval (5.8), there are many possible choices for u_1. The band of initial states for which the optimal control is not unique is the region marked A in Fig. 5.1. Points lying along a line with constant X_2 within this band are all controllable to the origin in the same time t_1, and the controllable set is not strictly convex. If X_2 is positive, points in the band B have non-unique optimal controls with the optimal time $t_1 = X_2$.

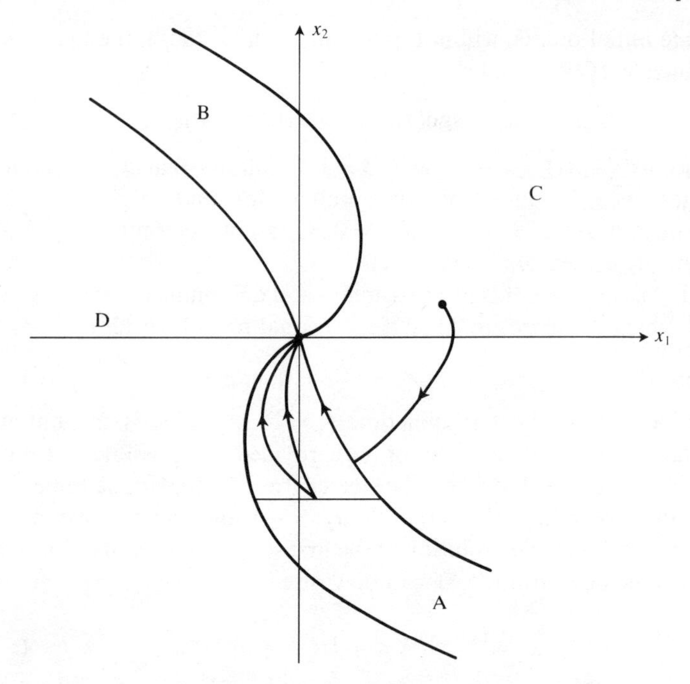

Fig. 5.1. Initial sets lying in the regions A and B have non-unique optimal controls. The optimal controls for initial states in regions C and D are unique.

Suppose the initial state lies in region C of Fig. 5.1, so that $X_1 > \frac{1}{2}X_2^2 - X_2$ if $X_2 < 0$ and $X_1 > X_2 - \frac{1}{2}X_2^2$ if $X_2 > 0$. Now the controls are uniquely determined and, from (5.4), we select $\alpha < 0$ and $u_1 = -1$. (It is easy to verify that the opposite choice of sign for α is the appropriate one for an initial state in region D.) We can also see from (5.4) that u_2 has at most one change of sign and that this change must be from -1 to $+1$ and not the other way round. Since the boundary between regions A and C is the trajectory that uses controls $u_1 = -1, u_2 = +1$, and terminates at the origin, it is clear that the optimal strategy is to employ $u_1 = -1$, $u_2 = -1$ from the initial state until this boundary is reached and then to switch the value of u_2 to $+1$. The optimal trajectory for a typical point in region C is shown in Fig. 5.1. To determine the optimal time all we have to do is write down the equations of the two parts of the trajectory and ensure that they meet at the switching time. A simple calculation shows that, for a point in region C, the optimal time is given by

$$t_1 = X_2 - 2 + 2(\tfrac{1}{2}X_2^2 - X_2 + X_1 + 1)^{1/2}. \tag{5.9}$$

The corresponding solution for initial states in region D can be found in a similar fashion.

5.2. The steering problem

The forward motion of a ship is inherently unstable and, without the continuous application of a control, the ship would tend to move broadside on to its direction of motion. In this case, as in other similar examples of steering, the response of the system to the control is reduced by the presence of a resistance to the enforced change of direction. If we represent the deviation of the direction of the ship from its desired course by the state variable x_1, we can model the steering problem by the two state equations

$$\dot{x}_1 = x_2, \quad \dot{x}_2 = -qx_2 + u_1, \tag{5.10}$$

where q is a positive constant. These equations are similar to those for the positioning problem, but a resistance proportional to the velocity has been added. We suppose that initially the direction has a non-zero value X_1, which we can take to be positive, with x_2 equal to zero, and we wish to regain the desired state, $x_1 = 0$ and $x_2 = 0$, as quickly as possible. The control at our disposal applies an angular acceleration to the ship in either direction, with a maximum value of unity, so $|u_1| \leq 1$.

In the standard notation, the matrices **A** and **B** are given by

$$\mathbf{A} = \begin{bmatrix} 0 & 1 \\ 0 & -q \end{bmatrix}, \quad \mathbf{B} = \begin{bmatrix} 0 \\ 1 \end{bmatrix}, \tag{5.11}$$

and (compare Exercise 2.3(d))

$$\exp(-\mathbf{A}t) = \begin{bmatrix} 1 & [1 - \exp(qt)]/q \\ 0 & \exp(qt) \end{bmatrix}. \tag{5.12}$$

The maximum principle with $\mathbf{h} = (\alpha, \beta)^{\mathrm{T}}$ then gives

$$u_1 = \operatorname{sgn}[\alpha q^{-1} + (\beta - \alpha q^{-1})\exp(qt)] \tag{5.13}$$

so that u_1 has at most one change of sign.

It is clear that, from the chosen initial state, it is necessary to start with a negative value for the control. It is also clear that, as the target is approached, the sign of the control must reverse if the state variable x_2 is to become zero. For the first stage, with $u_1 = -1$, the trajectory is given by solving (5.10), and we find that

$$x_2 = \frac{\exp(-qt) - 1}{q}, \quad x_1 = X_1 + \frac{1 - qt - \exp(-qt)}{q^2}. \tag{5.14}$$

For the second stage, with $u_1 = 1$, the trajectory that reaches the origin at time t_1 is given by

$$x_2 = \frac{1 - \exp[q(t_1 - t)]}{q}, \quad x_1 = \frac{\exp[q(t_1 - t)] - 1 - q(t_1 - t)}{q^2}. \tag{5.15}$$

The two trajectories must meet at some time t_2, say. Equating the two values of x_2 at the switching time gives the equation

$$2\exp(qt_2) = \exp(qt_1) + 1, \tag{5.16}$$

and if we equate the two values of x_1 at this time, we find that

$$qX_1 = 2t_2 - t_1. \tag{5.17}$$

Thus the optimal time t_1 can be found by solving the equation

$$q^2 X_1 = 2\ln\left(\frac{\exp(qt_1) + 1}{2}\right) - qt_1, \tag{5.18}$$

which can be rearranged to give

$$\exp(\tfrac{1}{2}q^2 X) = \cosh(\tfrac{1}{2}qt_1), \tag{5.19}$$

so that

$$t_1 = (2/q)\ln\{\exp(\tfrac{1}{2}q^2 X) + [\exp(q^2 X) - 1]^{1/2}\}. \tag{5.20}$$

For small values of qt_1 we can expand the right-hand side of (5.20) as a Taylor series. We then find that, provided $q^2 X_1 \ll 1$,

$$t_1 \approx 2X_1{}^{1/2} + \tfrac{1}{12}q^2 X_1{}^{3/2}. \tag{5.21}$$

As $q \to 0$ we recover the solution of the positioning problem with no damping, as in §4.4(2). For $q^2 X_1 \gg 1$ the asymptotic value of t_1 is given by

$$t_1 \approx qX_1 + \frac{2\ln 2}{q}. \tag{5.22}$$

The switching time t_2 in the first limiting case is approximately $\frac{1}{2}t_1$, so the two controls are applied for nearly equal times. In the second limiting case, however,

$$t_2 \approx qX_1 + \frac{\ln 2}{q}, \tag{5.23}$$

so that the control $u_1 = -1$ is applied for nearly the whole of the optimal time. Since q is very large, the value of x_1 is changing very slowly; when it is very near zero, application of the reverse control $u_1 = 1$ for a short time can bring the system to rest.

5.3. The harmonic oscillator

In a wide variety of dynamical problems, the situation occurs in which the system oscillates about a position of equilibrium. A displacement from the equilibrium position introduces a restoring force which moves the system towards equilibrium, but it overshoots and a continuing oscillatory motion ensues. In most practical situations there will be some damping in the system which will eventually reduce the amplitude of the oscillations. But this may take a considerable time, and it may be necessary to introduce some controlling mechanism to reduce the period during which the system is not in equilibrium. A familiar example of this type of motion is provided by a heavy weight suspended from a support and swinging in a vertical plane. A horizontal force can be applied to the weight in an attempt to bring the weight to rest. For a force of a given maximum size, the question is to determine what force should be used in order to bring the weight to rest in the shortest possible time. It is intuitively obvious that the force should be bang-bang, that is, we should always use a force of the maximum available size, in one direction or the other. When we should switch the direction of the force, however, is not at all obvious. We might guess that the best strategy would be to apply the force in the opposite direction to that in which the weight is moving. When we make use of the theory we have developed for solving this type of problem, however, we discover that this is *not* the optimal strategy.

If we represent the displacement of the object from its equilibrium position by x_1 and its velocity by x_2, the state equations, reduced to their simplest form by a suitable normalization, can be written as

$$\dot{x}_1 = x_2, \quad \dot{x}_2 = -x_1 + u_1, \tag{5.24}$$

where the control u_1 must lie between -1 and $+1$. The initial state has the aribitrary value (X_1, X_2) and the objective is to reach the target $(0, 0)$ in the shortest possible time t_1. The matrix $\mathbf{A}$ for this problem is given by

$$\mathbf{A} = \begin{bmatrix} 0 & 1 \\ -1 & 0 \end{bmatrix}, \tag{5.25}$$

and, from (2.17) with $k = 1$ and the sign of t changed,

$$\exp(-\mathbf{A}t) = \begin{bmatrix} \cos t & -\sin t \\ \sin t & \cos t \end{bmatrix}. \tag{5.26}$$

With $\mathbf{h}^{\mathrm{T}} = (\alpha, \beta)$, the maximum principle gives

$$u_1 = \operatorname{sgn}(\beta \cos t - \alpha \sin t) = \operatorname{sgn}(\delta \cos(t + \varepsilon)), \tag{5.27}$$

if $\delta \cos \varepsilon = \beta$ and $\delta \sin \varepsilon = \alpha$. It is at once clear that this problem differs from all those we have so far encountered. There is no bound to the number of zeros of the function $\cos(t + \varepsilon)$, so we cannot be sure how many switches the optimal control may have to undergo. It will be part of our task to determine the number of switches required for a given initial state. As we anticipated, (5.27) ensures that we only employ controls at full strength, but it might seem that the application of TOP supplies insufficient information to enable us to proceed, especially as we have no immediate way to determine the constants δ and ε. There is, however, one further item that we can deduce from (5.27) which, as we shall see, enables us to complete the solution. Although we do not know, from (5.27) alone, exactly when the switches occur, we do know that they must occur at intervals of π. The first switch can occur at any time not greater than π after the first application of a control. The second switch, if needed, cannot occur before a time π has elapsed since the first switch, and it must occur then. Similarly, if after a further time π has elapsed the target has still not been reached, a third switch must occur then, and so on.

From the state equations (5.24), it follows that

$$(x_1 - u_1)\dot{x}_1 + x_2\dot{x}_2 = 0, \tag{5.28}$$

so that when u_1 is held constant the trajectory is part of the curve

$$(x_1 - u_1)^2 + x_2^2 = c^2, \tag{5.29}$$

for some constant c. Thus the optimal trajectory is made up of arcs of circles, with centre at $(1, 0)$ when $u_1 = 1$ and centre at $(-1, 0)$ when $u_1 = -1$. If we write

$$x_1 - u_1 = c \cos \theta, \quad x_2 = c \sin \theta, \tag{5.30}$$

and substitute these forms into (5.24), we find that $\dot{\theta} = -1$. Hence the circular paths are traced out in the clockwise direction and the time taken to move between two points on the same arc is equal to the angle between the two radii joining them to the centre of the arc. It follows that, in the optimal solution, the maximum amount of any one circular arc that can be employed is a semicircle, and when that has been used, the centre of the next arc must switch to the alternative position.

Instead of starting at the initial state and working forwards in time, it is more convenient to work backwards from the target and to consider all the points from which the target can be reached by controls with an increasing number of switches. The origin lies on the circle with centre $(1, 0)$ and radius 1, so that it can be approached from negative values of x_2 (since the path is traversed in the clockwise direction) using the control $u_1 = 1$. We know that not more than a semicircle can be used without switching, so the points lying on this circle below the x_1 axis can be controlled by this choice of u_1. Let C_n be the semicircle below the x_1 axis, with radius 1 and centre at $(n, 0)$, and let C_{-n} lie above the x_1 axis and have centre at $(-n, 0)$ and radius 1. Points on C_1 can be controlled by $u_1 = 1$ without any switch, and, by symmetry, points lying on C_{-1} can be controlled by $u_1 = -1$. The final approach to the target for the optimal trajectories from any initial states must be along one or other of these two semicircles. For definiteness, let us concentrate on those paths whose last sections lie on C_1. To reach C_1 we must be using the opposite control, $u_1 = -1$, so the path will be an arc of a circle with centre B, with coordinates $(-1, 0)$. The maximum possible length of arc is a semicircle, so that the end of the arc through a point P on C_1 will be the opposite end of the diameter of the circle with centre B through P. These end points will lie on the reflection of C_1 in B, that is, they will lie on C_{-3}. Hence all points in the region S_1 of Fig. 5.2 can be controlled to the origin by the control sequence $\{-1, 1\}$ with one switch. The boundaries of S_1 are C_1, C_{-1}, C_{-3}, and the semicircle with centre B and radius 3. The reflection of C_{-3} in A, coordinates $(1, 0)$, is C_5, and points lying in S_2 require the

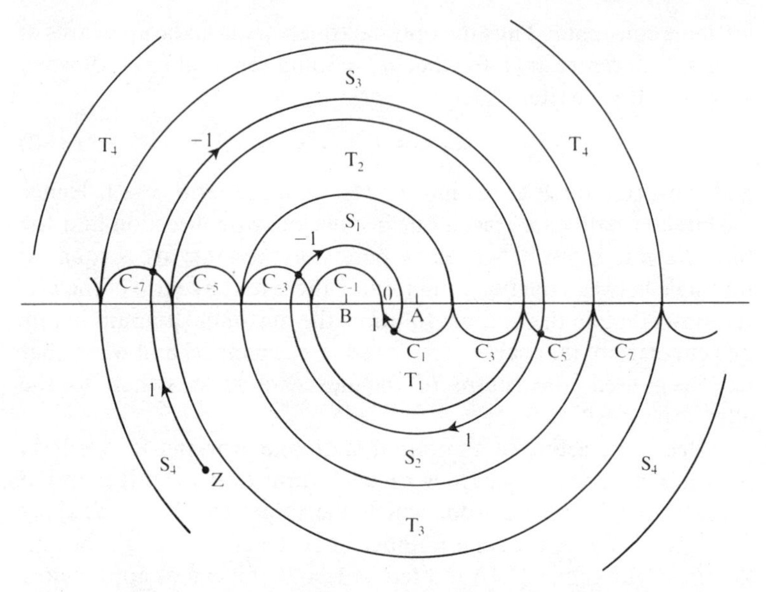

Fig. 5.2. Optimal trajectories for the harmonic oscillator.

control sequence $\{1, -1, 1\}$ to reach the origin optimally. The boundaries of S_2 are C_{-3}, C_5, and the semicircles with centre A and radii 3 and 5. Proceeding in this manner, we cover half the plane by regions of this type; for any point lying in one of these regions a sequence of controls ending with $+1$ must be used. The trajectory for a point Z in S_4 is also shown in Fig. 5.2. The other half of the plane is covered by regions T_1, T_2, etc. for which the optimal control sequence ends with -1.

It is now clear how we construct the optimal trajectory for any initial state. Note that the control -1 is needed when the trajectory lies in regions of either kind that lie mainly above the x_1 axis, and $+1$ when they are below, so that the centre of the initial arc is known. Draw this arc until it reaches C_n for some n; switch the centre of the circle to the opposite point, A or B, and draw a semicircle, which will end on another one of the switching semicircles C_n. Repeat until either C_1 or C_{-1} is reached when a final switch will take the point to the origin. The optimal time can be calculated by adding up the angular distances along each section of the path. For the point Z in Fig. 5.2, the optimal time is approximately $\frac{1}{4}\pi + 3\pi + \frac{1}{3}\pi = 3.58\pi$.

The intuitive strategy suggested before we worked out the optimal solution was to switch the control whenever the direction of the motion changes, that is, whenever the trajectory crosses the x_1 axis. We have found that the optimal solution requires the switching to take place on the semicircles C_n. Of course, for initial states that are a long distance from the origin, the relative difference between the two costs will be small.

5.4. A predator–prey problem

Population counts made in McKinley National Park in Alaska of lynx and of snow-shoe hares have shown that the numbers of both species have a cyclic variation with a period of several years. There is a 90° phase difference between them, with extremely high or low numbers for one species coinciding with mean values for the population of the other. This variation can be accounted for by the fact that the main source of food for the lynx is the snow-shoe hares, and that when their prey is plentiful their numbers increase, which in turn gives rise to a depletion of the number of hares and a subsequent decrease in the lynx population. The environment in the National Park is free from external control by hunting or trapping. A constant population for both species is possible, but any disturbance leads to a continuing cyclic variation, as in the comparable problem of the harmonic oscillator we discussed in §5.3.

In similar circumstances in other locations, for a variety of reasons it may be desirable to remove the cyclic variation and return the system to its stable steady state. One way in which this might be attempted would be to introduce or remove members of either species, but the timing of such intervention is crucial to the achievement of the desired goal. For definiteness, let us suppose that the only strategy open to us is to introduce additional predators at a given maximum rate. With this limitation on the available control, is it possible to regain constant populations from any initial state? If so, when and for how long should the control be applied in order to reach equilibrium in the shortest possible time?

Suppose the population of predators at time t is given by $y_1(t)$ and that of their prey by $y_2(t)$, the equilibrium values being Y_1 and Y_2, respectively. Then model equations for the evolution of any departure from equilibrium can be written as

$$\dot{y}_1 = y_2 - Y_2, \quad \dot{y}_2 = -(y_1 - Y_1). \tag{5.31}$$

These equations state that the number of predators increases when there is an excess of their prey and that the number of the prey decreases when there are too many predators. For simplicity, and without concealing any significant features of the system, we have set all parameters equal to unity. If, at time $t = 0$, $y_1 = Y_1$ and $y_2 = Y_2 + c$, there is an initial abundance or depletion of the prey, depending on the sign of c. The solution of the state equations (5.31) is easily found to have the form

$$y_1 = Y_1 + c \sin t, \quad y_2 = Y_2 + c \cos t, \tag{5.32}$$

which indicates the periodic property of the solution. The initial state is regained after a time equal to 2π. The linear form of the state equations is only justified if the departures from equilibrium are relatively small; otherwise we might deduce the appearance of negative populations. Now let us consider how this same initial condition can lead to the equilibrium state being restored by releasing additional predators into the system at a rate $v_1(t)$, say, where v_1 can take any value in the range $0 \leq v_1 \leq 2$. The state equations now have the form

$$\dot{y}_1 = y_2 - Y_2 + v_1, \quad \dot{y}_2 = -y_1 + Y_1. \tag{5.33}$$

If we define new state and control variables by

$$x_1 = y_1 - Y_1, \quad x_2 = 1 + y_2 - Y_2, \quad u_1 = v_1 - 1, \tag{5.34}$$

the state equations become

$$\dot{x}_1 = x_2 + u_1, \quad \dot{x}_2 = -x_1, \tag{5.35}$$

where $-1 \leq u_1 \leq +1$ and the initial state is given by $x_1 = 0$, $x_2 = 1 + c$. The equilibrium state is now given by $x_1 = 0$, $x_2 = 1$, and we wish to reach this state at time t_1 with t_1 as small as possible. When it is reached, we can maintain the system there by imposing the control $u_1 = -1$, which means, of course, that v_1 is zero and we stop interfering with nature.

The state equations (5.35) are very similar to those of the harmonic oscillator in §5.3. With $\mathbf{h}^{\mathrm{T}} = (\alpha, \beta)$ the maximum principle TOP shows that

$$u_1 = \operatorname{sgn}(\alpha \cos t + \beta \sin t), \tag{5.36}$$

so that u_1 switches between its extreme values of $+1$ and -1 at intervals equal to π. It is clear that, as t approaches the terminal time

t_1, we must use the control $+1$. For, if $u_1 = -1$, the trajectory through the target point is the target point $x_1 = 0$, $x_2 = 1$ itself and we would have reached the target at an earlier time. Hence t_1 would not be the optimal solution. Let A_n be the point $(0, n)$. The trajectories for which $u_1 = +1$ are circular arcs with centre A_{-1} and those with $u_1 = -1$ are circular arcs with centre A_1. Thus the final approach to the target is along the trajectory with equation

$$x_1^2 + (x_2 + 1)^2 = 4. \tag{5.37}$$

As in the solution for the harmonic oscillator in §5.3, we can work backwards in time from this trajectory, remembering that no control can be used for a period longer than π and that a switch cannot occur until this time has elapsed after the previous switch. In this way we can construct a diagram similar to that in Fig. 5.2; the difference is that the target point is no longer the origin. Part of such a diagram is shown in Fig. 5.3.

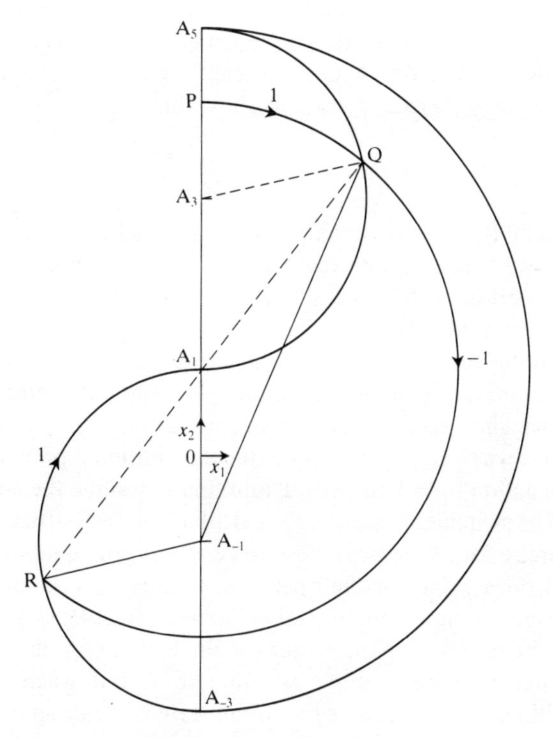

Fig. 5.3. An optimal trajectory for the predator–prey equation.

The points for which the control sequence $\{-1, 1\}$ is required are those that lie in the region bounded by the semicircular arcs with centre A_3 and radius 2 and centre A_1 and radius 4 with x_1 positive and by the semicircular arc with centre A_{-1} and radius 2 with x_1 negative. For definiteness, let us consider the optimal trajectory from the initial state with $0 < c < 4$. From the initial point P, with coordinates $(0, 1 + c)$, we apply the control $u_1 = 1$ until we reach the point Q on the semicircle with centre A_3. The trajectory from P to Q is an arc of a circle centre A_{-1} and radius $2 + c$. With control $u_1 = -1$ we move from Q to R along the semicircular arc with centre A_1, and then we use $u_1 = 1$ to move from R to A_1 along the final trajectory (5.37). Since A_1 is the midpoint of $A_{-1}A_3$ and of QR, $A_{-1}RA_3Q$ is a parallelogram. The optimal time is the total angle turned through on the three parts of the trajectory and so is the sum of the angle between $A_{-1}P$ and $A_{-1}Q$, π, and the angle between $A_{-1}R$ and $A_{-1}A_1$. The total angle is thus equal to π plus the angle between $A_{-1}R$ and $A_{-1}Q$, and because $A_{-1}R$ and QA_3 are parallel this is equal to 2π minus the angle between $A_{-1}Q$ and A_3Q. The sides of the triangle $A_{-1}A_3Q$ have lengths equal to 2, 4, and $2 + c$, and the cosine rule can be used to show that

$$t_1 = 2\pi - \cos^{-1}\left(\frac{c^2 + 4c - 8}{4c + 8}\right). \tag{5.38}$$

The optimal time lies between π and 2π when c lies between 0 and 4, so the steady state can be regained in a time no greater than one complete period of the natural cycle. With $c = 4$, we take no action until a time π has elapsed, which is half the period of the natural cycle, and the number of predators has reached a minimum. We then introduce predators at the maximum rate and, after another half-period, we shall have reached the equilibrium state. For smaller values of c, we must begin by introducing additional predators, then stop this action for a half-period, and finally resume the controlling action. The switching times can be calculated for any given value of c (see Exercise 5.4). When c is close to zero, but not equal to zero, the first and third stages of the operation occupy very small intervals and the optimal time is mainly taken up by the middle stage when no action is being taken. Notice that the limit of the optimal time as c approaches zero from above is equal to π, but when $c = 0$ the optimal time t_1 is zero, since the initial state is the desired equilibrium one.

The initial state of the system contains an abundance of prey and, as might be expected, the optimal strategy begins by introducing predators. However, it is essential to end this process at a time when the prey is still abundant. Similar arguments to those presented here can be applied to find the solution for other values of c and for other options; for example, we could control the system by adding to the size of the population of the prey instead of that of the predators or by culling the predators (see Exercise 5.5).

In Part A we have proved the existence of a solution of the linear time-optimal control problem and established TOP, a maximum principle that enables the solution of such problems to be constructed. The examples in this chapter have shown how TOP can be applied to a variety of situations where it is desired to control a system to a specified target in the shortest possible time. In each case the admissible controls have been bounded and the optimal solutions have all required the use of bang–bang controls. In Part B we extend the range of optimal control problems to a wider class of cost functions, some of which may allow unbounded controls, and to nonlinear state equations.

Exercises 5

(*Solutions on p. 224*)

5.1 The state equations for a two-dimensional system with two controls are

$$\dot{x}_1 = -x_1 + u_1, \quad \dot{x}_2 = u_1 + u_2,$$

and $|u_1| \le 1, |u_2| \le 1$. The target is $x_1 = 0, x_2 = 0$, and the initial state is $x_1 = X_1, x_2 = X_2$ with X_2 positive. Show that the optimal controls are not uniquely determined if

$$X_1 > \exp(\tfrac{1}{2}X_2) - 1$$

and find the optimal time in this case.

Find the optimal controls for values of X_1 for which this inequality is not satisfied (with the same value of X_2) and determine the optimal time.

5.2 The target in the harmonic-oscillator problem of §5.3 was the origin. Suppose that it is desired to bring the oscillator to rest, not necessarily at the origin, so the target set is $x_2 = 0$. Find the optimal control and the optimal time for an initial state $x_1 = X_1, x_2 = X_2$ with X_2 positive.

5.3 The state equations for the harmonic oscillator with two controls are given by

$$\dot{x}_1 = x_2 + u_1, \quad \dot{x}_2 = -x_1 + u_2.$$

Both controls belong to $\mathscr{U}_b$, and the target is the origin.

Show that the pairs of controls in the optimal solution must go through the sequence $\{(1, 1), (1, -1), (-1, -1), (-1, 1)\}$, repeated if necessary, and that the switchings from one pair of controls to the next must take place at intervals of $\frac{1}{2}\pi$.

Find the optimal trajectory and the optimal time for the initial state $x_1 = 1, x_2 = 3$.

5.4 Calculate the switching times for the predator–prey example worked out in §5.4, for which the initial state was $x_1 = 0$, $x_2 = 1 + c$, with $0 < c < 4$.

5.5 Find the solution of the predator–prey problem when the control option is that of culling the predators, that is, $0 \geq v_1 \geq -2$. Consider the same initial state as before, namely $y_1 = Y_1$, $y_2 = Y_2 + c$, with $0 \leq c \leq 4$. Compare the optimal time for this option with that found in §5.4 for the alternative option with v_1 positive.

5.6 The population of a pest is increasing exponentially. To control the pest, a predatory beetle is developed and introduced into the environment. The beetles are sterile, so they will eventually die off if the supply is terminated, but since they are themselves harmful to the crop infested by the pest it is desirable that they should be removed as soon as possible. Let the populations of the pest and the beetles be denoted by $x_1(t)$ and $x_2(t)$ per hectare, respectively. Initially $x_1 = X$, $x_2 = 0$, and the target is to reduce both populations to zero simultaneously and in the shortest possible time. Model equations for the two populations can be taken as

$$\dot{x}_1 = x_1 - x_2, \quad \dot{x}_2 = -x_2 + u_1, \quad -1 \leq u_1 \leq 1,$$

where u_1 represents the rate at which the beetles are introduced or eliminated.

Show that the given initial state can only be controlled if $X < \frac{1}{2}$, and explain why this is so. Find the optimal controls and optimal time when $X < \frac{1}{2}$. Show that the optimal time tends to infinity as X tends to $\frac{1}{2}$.

PART B
The Pontryagin Maximum Principle

The Pontryagin maximum principle (PMP) applies to a much wider class of optimal control problems than that considered in Part A. There is no restriction to linear state equations and many different cost functions can be successfully handled. Since the problems solved in Part A are a subset of those covered by the PMP, the inclusion of Part A in this book might be questioned. There are, however, some respects in which the contents of Part A are superior. The proof of the PMP is quite difficult and will only be given in outline. Also, the PMP only provides *necessary* conditions that an optimal solution must satisfy if it exists. There is, in general, no guarantee that a given problem will have an optimal solution. In Part A we were able to prove the existence of a solution.

6 The basic Pontryagin maximum principle

Wisdom denotes the pursuing of the best ends by the best means.

Francis Hutcheson

In this chapter we state the Pontryagin maximum principle (PMP) in its simplest form and use it to solve some simple examples. Extensions to a less restricted class of problems are discussed in Chapter 7, but the proof of the PMP is postponed to Chapter 9.

6.1. An illustrative example

Before giving the complete statement of the PMP, and in order to see clearly how the principle is applied, we look first at a very simple problem in detail and go through the steps required in using the PMP to obtain a solution. A system in unstable equilibrium is one in which any deviation from equilibrium increases if the system is left to itself. The application of a control aims to reduce the deviation to zero, but at some cost depending on the size of the control that has to be employed. Suppose the state of the system is given by a single state variable x_1, with $x_1 = 0$ in equilibrium. If u_1 is the control, a state equation that typifies potentially unstable behaviour is

$$\dot{x}_1 = x_1 + u_1. \tag{6.1}$$

Without any control, x_1 would increase exponentially from an initial value of $X > 0$. The target to be reached in the given time t_1 is the equilibrium state. Thus the initial and final conditions are

$$x_1(0) = x_1^0 = X, \quad x_1(t_1) = x_1^1 = 0. \tag{6.2}$$

We select as the cost function one that penalizes the use of a control with a large magnitude and define the cost J by

$$J = \int_0^{t_1} \tfrac{1}{2}u_1^2 \, \mathrm{d}t. \tag{6.3}$$

The control u_1 is to be a piecewise-continuous function of t, and we do not set any bounds on its size.

The problem is now completely defined; in terms of the classification introduced in Chapter 1, it is a one-dimensional linear autonomous problem, with one control variable without bounds and with a fixed target and a fixed terminal time. We now go through the procedure of applying the PMP.

The first step is to replace the cost integral by introducing an additional state variable x_0 which satisfies the state equation

$$\dot{x}_0 = \tfrac{1}{2}u_1^2, \tag{6.4}$$

with boundary conditions

$$x_0^0 = 0 \quad \text{and} \quad x_0^1 = J. \tag{6.5}$$

The second step is to introduce two *co-state variables* z_0 and z_1, and to form the quantity $H(x_0, x_1, z_0, z_1, u_1)$ defined by

$$H = z_0\dot{x}_0 + z_1\dot{x}_1 = \tfrac{1}{2}z_0u_1^2 + z_1(x_1 + u_1). \tag{6.6}$$

In analytical dynamics a corresponding expression is called the Hamiltonian of the system. From this Hamiltonian the rates of change of the variables are prescribed by Hamilton's equations. In a similar manner, we state here that z_0 and z_1 must satisfy equations of Hamiltonian form, namely

$$\dot{z}_0 = -\frac{\partial H}{\partial x_0} = 0, \quad \dot{z}_1 = -\frac{\partial H}{\partial x_1} = -z_1. \tag{6.7}$$

The other pair of Hamilton's equations are just the state equations

$$\dot{x}_0 = \frac{\partial H}{\partial z_0} = \tfrac{1}{2}u_1^2, \quad \dot{x}_1 = \frac{\partial H}{\partial z_1} = x_1 + u_1. \tag{6.8}$$

The third step is to solve the co-state equations, which will introduce some constants whose values are yet to be determined. The equation for $\dot{z}_0$ shows that $z_0 =$ constant, and the PMP requires that this constant should be *negative*; without loss of generality we can choose

$$z_0 = -1. \tag{6.9}$$

The equation for $\dot{z}_1$ has the solution

$$z_1 = Ae^{-t}, \tag{6.10}$$

where A is a constant which we cannot determine at this stage.

Now comes the fourth and crucial step in applying the PMP. We choose the control $u_1(t)$ to give the greatest possible value for the Hamiltonian H, considered as a function of u_1. Since we can write H in the form

$$H = -\tfrac{1}{2}(u_1 - z_1)^2 + z_1 x_1 + \tfrac{1}{2}z_1^2, \tag{6.11}$$

we see that we must have

$$u_1 = z_1 = A\mathrm{e}^{-t}. \tag{6.12}$$

We have now determined the control function that has to be used, apart from the constant A. But we still have to ensure that x_1 moves from its initial to its final value under the application of this control. The fifth step is to solve the state equations for x_0 and x_1. The equation for $\dot{x}_1$ now has the form

$$\dot{x}_1 = x_1 + A\mathrm{e}^{-t} \tag{6.13}$$

and the solution that satisfies the initial condition is

$$x_1 = -\tfrac{1}{2}A\mathrm{e}^{-t} + (X + \tfrac{1}{2}A)\mathrm{e}^{t}. \tag{6.14}$$

The target $x_1 = 0$ is to be reached at time t_1, so that

$$-\tfrac{1}{2}A\mathrm{e}^{-t_1} + (X + \tfrac{1}{2}A)\mathrm{e}^{t_1} = 0 \tag{6.15}$$

and hence

$$A = -\frac{2X}{1 - \exp(-2t_1)}. \tag{6.16}$$

We have now determined the optimal control and the optimal trajectory. The optimal cost comes from the state equation for $\dot{x}_0$ and we find that

$$\begin{aligned} J &= \int_0^{t_1} \tfrac{1}{2}u_1^2\,\mathrm{d}t = \int_0^{t_1} \tfrac{1}{2}A^2 \exp(-2t)\,\mathrm{d}t \\ &= \frac{X^2}{1 - \exp(-2t_1)}. \end{aligned} \tag{6.17}$$

This completes the solution of the problem; provided the problem has an optimal solution, this procedure has determined it, and the solution is unique.

There is one more result which forms part of the PMP but which we have not needed to use. If we evaluate the Hamiltonian H along the optimal trajectory, that is, with the state, co-state, and control variables given their optimal values, we find that at time t

$$H = z_1(x_1 + \tfrac{1}{2}z_1) = Ae^{-t}(X + \tfrac{1}{2}A)e^t = A(X + \tfrac{1}{2}A), \quad (6.18)$$

which is a constant. This is not an isolated result for this particular problem; the PMP states that the Hamiltonian is always constant along the optimal solution.

There is no obvious way by which we could have predicted that the solution of this problem would have the form we have found. It is not surprising that the control is always negative, since we start with x_1 positive and we wish to decrease it. Also, since the strength of the pull away from equilibrium decreases as x_1 decreases, we could have anticipated that the control would diminish as the system approaches equilibrium. But the precise form of the control to be employed, and the optimal cost, cannot be determined except by application of the PMP or by some equivalent procedure.

There is something further that can be learnt from this example. The optimal cost is a decreasing function of the terminal time t_1. If we allow the system a longer time to reach equilibrium, we reduce the cost; even though the integral for the cost is over a longer range, the decrease in the value of the control is dominant and the cost decreases. However large t_1 is, the cost is always greater than X^2. It follows that, if we had not imposed a fixed terminal time but allowed it to take any finite value, there would be *no* optimal solution. The cost could be made as close as we please to the value X^2, but it could never actually attain this value. This typifies the situation in which a problem that seems to be satisfactorily posed does not have an optimal solution. Of course, this is a mathematical difficulty rather than a practical one. Since a cost very close to the absolute minimum can be achieved, the failure to achieve this minimum exactly is of no importance.

6.2. Statement of the PMP in its basic form

Now that we have seen how to apply the PMP to a particular problem, we can proceed to give the formal statement of the principle in its basic form.

We consider an autonomous problem, linear or nonlinear, with a fixed target and no terminal cost, as defined in Chapter 1. The state vector $\mathbf{x}$ is n-dimensional and the state equations are

$$\dot{\mathbf{x}} = \mathbf{f}(\mathbf{x}, \mathbf{u}), \tag{6.19}$$

where $\mathbf{u}$ is an m-dimensional control vector. The initial state of the system at $t = 0$ is given by $\mathbf{x}^0$ and the terminal state at $t = t_1$ is $\mathbf{x}^1$. The terminal time t_1 is either *fixed* or *free*. The control $\mathbf{u}$ has components u_i, $i = 1, 2, \ldots, m$, which are piecewise-continuous functions of t and may either be without bounds or satisfy a restriction to values in the range from -1 to $+1$, that is,

$$|u_i(t)| \leq 1, \quad \text{for } 0 \leq t \leq t_1 \quad \text{and } i = 1, 2, \ldots, m. \tag{6.20}$$

We refer to these two sets of controls as $\mathscr{U}_\mathrm{u}$ and $\mathscr{U}_\mathrm{b}$ for the unbounded and bounded cases, respectively. In either case, we shall say that the control $\mathbf{u}$ belongs to the set $\mathscr{U}$ of all admissible controls, which can stand for either of these two sets, as specified in any particular problem.

The cost is defined by

$$J = \int_0^{t_1} f_0(\mathbf{x}, \mathbf{u})\, \mathrm{d}t \tag{6.21}$$

and we seek the optimal control $\mathbf{u} \in \mathscr{U}$ such that the state of the system can be steered from $\mathbf{x}^0$ to $\mathbf{x}^1$ in time t_1 with the minimum value of J. The trajectory $\mathbf{x}(t)$ using the optimal control is the optimal trajectory, and the corresponding values of J and t_1 (if free) are the optimal cost and time, respectively.

We first define an extra state variable x_0 by the state equation

$$\dot{x}_0 = f_0(\mathbf{x}, \mathbf{u}), \quad x_0(0) = 0, \tag{6.22}$$

so that the cost is given by

$$J = x_0(t_1). \tag{6.23}$$

We next introduce the *extended state vector* $\hat{\mathbf{x}}$, of dimension $n + 1$, whose components are x_i, $i = 0, 1, \ldots, n$. If we define the extended vector $\hat{\mathbf{f}}$ similarly, the state equations can be written in the form

$$\dot{\hat{\mathbf{x}}} = \hat{\mathbf{f}}(\mathbf{x}, \mathbf{u}). \tag{6.24}$$

The Hamiltonian $H(\mathbf{x}, \hat{\mathbf{z}}, \mathbf{u})$ is defined by

$$H = \hat{\mathbf{z}}^\mathrm{T}\dot{\hat{\mathbf{x}}} = \sum_{i=0}^{n} z_i f_i, \tag{6.25}$$

where $\hat{\mathbf{z}}$ is the *extended co-state vector* of dimension $n + 1$. Hamilton's equations then give the state equations (6.24) and the co-state equations

$$\dot{\hat{\mathbf{z}}} = -\frac{\partial H}{\partial \hat{\mathbf{x}}}. \tag{6.26}$$

Since H does not depend on x_0, these equations can be written in the form

$$\dot{z}_0 = 0, \quad \dot{z}_i = -\frac{\partial H}{\partial x_i}, \quad i = 1, 2, \ldots, n. \tag{6.27}$$

The PMP can now be stated as follows.

> Suppose that the problem we have defined has an optimal solution with an optimal control $\mathbf{u}^*$. Then the following conditions must hold:
>
> (i) $z_0 = -1$;
>
> (ii) $\mathbf{u}^*$ is the control function for which $H(\mathbf{x}, \hat{\mathbf{z}}, \mathbf{u})$ has its supremum for all $\mathbf{u}$ belonging to $\mathscr{U}$, the set of admissible controls;
>
> (iii) the co-state equations have a solution $\hat{\mathbf{z}}^*$, and the state equations a solution $\mathbf{x}^*$ which takes the values $\mathbf{x}^0$ at $t = 0$ and $\mathbf{x}^1$ at $t = t_1$;
>
> (iv) the Hamiltonian is constant along the optimal trajectory and this constant is zero if the terminal time is free, that is,
>
> $$\begin{aligned} H(\mathbf{x}^*, \hat{\mathbf{z}}^*, \mathbf{u}^*) &= \text{constant}, \quad \text{if } t_1 \text{ is fixed,} \\ &= 0, \quad \text{if } t_1 \text{ is free and positive.} \end{aligned}$$

The $2n + 2$ state and co-state equations require an equal number of constants to fix their solution. These are provided by the n initial and n final conditions on $\mathbf{x}$, the known initial value of x_0, and the value of z_0, which is provided by the PMP itself. If t_1 is free, we need an extra condition to determine it, and this is provided by condition (iv) of the PMP.

It should be emphasized that the PMP provides *necessary* conditions that must be satisfied by an optimal solution, if one exists. There may be more than one solution that satisfies all the conditions of the PMP. In that case, we must choose the one that entails the least value of the cost J; we may have to accept the fact that there is more than one solution with the same least value of the cost, since there is no guarantee that the optimal solution is unique.

On the other hand, it may be impossible to satisfy all the conditions of the PMP, in which case we know that there is no optimal solution. We have already encountered this situation in the discussion after the solution of the illustrative example in §6.1. There we considered the free-time problem by examining the optimal costs for arbitrary values of the terminal time t_1, and found that their minimum could not be achieved. If we apply the PMP directly to the same problem with t_1 free, the solution follows as before except that now we must apply condition (iv) to fix the optimal time. The value of H along the optimal trajectory is given by (6.18), and condition (iv) states that this must be zero. Hence we must either have $A = 0$, which is impossible since then no control is applied, or $X + \frac{1}{2}A = 0$, which, combined with (6.16), shows that

$$1 - \exp(-2t_1) = 1. \tag{6.28}$$

Since there is no finite value of t_1 that satisfies this equation, we conclude that there is no optimal solution.

It may seem a little arbitrary that condition (i) of the PMP requires z_0 to be negative. This is indeed only conventional. We could have insisted that z_0 be positive; then condition (ii) would have required us to take the infimum of H instead of the supremum and we would have a Pontryagin minimum principle.

6.3. Some examples

We can now apply the PMP to some simple examples to demonstrate its versatility. Although the principle holds for nonlinear problems, their solution usually requires numerical methods and a computer. Since we are interested at present in showing how the principle can be applied, it is best to confine ourselves to problems for which the solution can be found in closed form. To start with, let us consider the problem discussed by elementary methods in §1.2.

1. *Plant growth*

The state equation for this problem is

$$\dot{x}_1 = 1 + u_1 \tag{6.29}$$

and the cost function in its differentiated form is

$$\dot{x}_0 = \tfrac{1}{2}u_1^2.$$

The terminal time $t_1 = 1$ and so is fixed, and the initial and final states are

$$x_1(0) = 0, \quad x_1(1) = 2. \tag{6.30}$$

The Hamiltonian is

$$H = \tfrac{1}{2}z_0 u_1^2 + z_1(1 + u_1) \tag{6.31}$$

and the co-state equations are

$$\dot{z}_0 = 0, \quad \dot{z}_1 = 0, \tag{6.32}$$

which have the solutions

$$z_0 = -1, \quad z_1 = A, \tag{6.33}$$

where A is a constant and we have made use of condition (i) of the PMP to fix the value of z_0. The control $u_1 \in \mathscr{U}_u$ and the Hamiltonian (6.31) has its largest value if we choose $u_1 = z_1$, as can be seen by completing the square as in (6.11) or by differentiating H with respect to u_1. The state equation then becomes

$$\dot{x}_1 = 1 + A, \tag{6.34}$$

and the solution that satisfies the initial and final conditions is

$$x_1 = 2t, \quad \text{with } A = 1. \tag{6.35}$$

The optimal control is therefore $u_1 = 1$ for all t and the optimal cost is $J = \frac{1}{2}$, as we found in §1.2 by elementary methods.

2. *Gout*

The disease of gout is characterized by an excess of uric acid in the blood. Drugs are available which will reduce the uric acid to an acceptable level. If we measure the excesss uric acid by the state variable x_1, a model equation that describes some features of this system has the form

$$\dot{x}_1 = -x_1 + 1 - u_1. \tag{6.36}$$

With $u_1 = 0$, that is, with no drugs, the state variable equilibrates at the level $x_1 = 1$, which is too high. The application of the drug aims to reduce the level to zero; in this model equation the drug is administered continuously and not in discrete doses. We can suppose that the initial state is the equilibrium one, so that $x_1(0) = 1$, and that at time t_1 the state is the acceptable one in which $x_1(t_1) = 0$.

It is clearly advantageous to attain the safe level of uric acid as quickly as possible, and this can be done by using large amounts of the drug. On the other hand, such large doses are inadvisable, both on medical grounds because of the side-effects they may induce, and on financial grounds since the drug is expensive. A cost function that balances these two components has the form

$$J = \int_0^{t_1} \tfrac{1}{2}(k^2 + u_1^2)\,\mathrm{d}t. \tag{6.37}$$

The first component increases with the terminal time and the second with the amount of drug utilized. The constant k measures the relative importance of the two components.

The Hamiltonian for this system has the form

$$H = -\tfrac{1}{2}k^2 - \tfrac{1}{2}u_1^2 + z_1(-x_1 + 1 - u_1), \tag{6.38}$$

where we have already inserted the value of z_0. The co-state equation is

$$\dot{z}_1 = z_1 \tag{6.39}$$

and, since $u_1 \in \mathscr{U}_u$, the supremum of H is attained by choosing $u_1 = -z_1$, so that we have

$$u_1 = -z_1 = Ae^t, \tag{6.40}$$

where A is a constant. This is a free-time problem, and we can apply condition (iv) at $t = 0$. Substituting the known values of the variables into (6.38) we obtain

$$-\tfrac{1}{2}k^2 - \tfrac{1}{2}A^2 - A(-1 + 1 - A) = 0, \tag{6.41}$$

and so

$$A = k. \tag{6.42}$$

Since the drug is being administered to the patient, it is clear that we require u_1 to be positive and we can ignore the other root $A = -k$.

We can now solve the state equation (6.36), making use of what we have already determined. The solution is

$$x_1 = 1 - \tfrac{1}{2}ke^t + \tfrac{1}{2}ke^{-t}, \tag{6.43}$$

where the initial condition has been applied. When the terminal condition is applied, we find that

$$1 - k \sinh t_1 = 0, \tag{6.44}$$

and the terminal time in the optimal solution is

$$t_1 = \sinh^{-1}(1/k). \tag{6.45}$$

The optimal cost can now be evaluated from (6.37) and we find that

$$\begin{aligned} J &= \tfrac{1}{2}k^2 t_1 + \tfrac{1}{4}k^2[\exp(2t_1) - 1] \\ &= \tfrac{1}{2}k^2 \sinh^{-1}(1/k) + \tfrac{1}{2}[1 + (k^2 + 1)^{1/2}]. \end{aligned} \tag{6.46}$$

When k is small, the prime consideration is to use a small amount of the drug, even if the treatment lasts for a long time. In the limit as $k \to 0$, $J \to 1$ but $t_1 \to \infty$, so there is no optimal solution with $k = 0$. If k is large, the important goal of the treatment is to reach the safe level as quickly as possible, with little concern about the dosage required. For large values of k, $t_1 \sim 1/k$ and $J \sim k$.

3. *The positioning problem*

The time-optimal problems solved in Part A can also be solved using the PMP. As an example, and so that the two methods of solution can be compared, we shall re-solve the positioning problem, which was formulated in §1.4 and solved in §4.4(2). We shall take the opportunity to extend the solution to cover arbitrary initial states. The state equations are

$$\dot{x}_1 = x_2, \quad \dot{x}_2 = u_1, \tag{6.47}$$

and the cost is equal to the time taken to reach the origin, so

$$\dot{x}_0 = 1 \tag{6.48}$$

and we have a free-time problem. The control u_1 is to be piecewise continuous, with bounds of -1 and $+1$, so that $u_1 \in \mathscr{U}_b$, and the initial and final states are

$$\begin{aligned} x_1(0) = X_1, \quad & x_2(0) = X_2, \\ x_1(t_1) = 0, \quad & x_2(t_1) = 0. \end{aligned} \tag{6.49}$$

The problem is to reach the target from the initial position in mimimum time.

The Hamiltonian can be written down immediately in the form

$$H = -1 + z_1 x_2 + z_2 u_1, \tag{6.50}$$

since $z_0 = -1$, as usual. The co-state equations are

$$\dot{z}_1 = 0, \quad \dot{z}_2 = -z_1, \tag{6.51}$$

and their solutions are

$$z_1 = A, \quad z_2 = B - At, \tag{6.52}$$

where A and B are constants. Since the control is bounded by -1 and $+1$, the supremum of H is attained when u_1 takes the value $+1$ when z_2 is positive, and the value -1 when z_2 is negative. Thus we have a bang–bang control, with the control switching between its extreme values when z_2 passes through zero. Since z_2 is a linear function of t, there can be at most one such switch.

It is convenient to trace the solution back from the final state. If the target is approached with $u_1 = +1$, the trajectory can be found from the state equations (6.47) and these give the results

$$x_2 = t - t_1, \quad x_1 = \tfrac{1}{2}(t - t_1)^2, \quad x_2^2 = 2x_1, \tag{6.53}$$

since the target is to be reached at $t = t_1$. Note that x_2 is negative along this trajectory. If X_2 is negative, and $X_2^2 = 2X_1$, the control $u_1 = +1$ can be applied without switching and we reach the target in the optimal time $t_1 = -X_2$. Otherwise, we must switch to $u_1 = -1$ at some time t_2, say, where t_2 must lie between zero and t_1. As no further switching is possible, this control must be applied for all t between 0 and t_2, and the solution that passes through the initial state is given by

$$\begin{gathered} x_2 = X_2 - t, \quad x_1 = X_1 + X_2 t - \tfrac{1}{2}t^2, \\ x_2^2 + 2x_1 = X_2^2 + 2X_1. \end{gathered} \tag{6.54}$$

The two trajectories (6.53) and (6.54) intersect where

$$4x_1 = X_2^2 + 2X_1, \quad x_2 = -(X_1 + \tfrac{1}{2}X_2^2)^{1/2} \tag{6.55}$$

and at $t = t_2$, so that

$$t_2 = X_2 + (X_1 + \tfrac{1}{2}X_2^2)^{1/2}, \tag{6.56}$$

and, from (6.53), the optimal time is given by

$$t_1 = t_2 - x_2 = X_2 + 2(X_1 + \tfrac{1}{2}X_2^2)^{1/2}. \tag{6.57}$$

This is the required solution, provided the square root is real and that t_2 is positive. These conditions are satisfied if

$$X_1 > -\tfrac{1}{2}X_2^2 \quad \text{when } X_2 > 0, \quad \text{and} \quad X_1 > \tfrac{1}{2}X_2^2 \quad \text{when } X_2 \le 0. \tag{6.58}$$

In Fig. 6.1, this means that the initial point must lie to the right of the switching curve through the origin. If it lies on either part of this

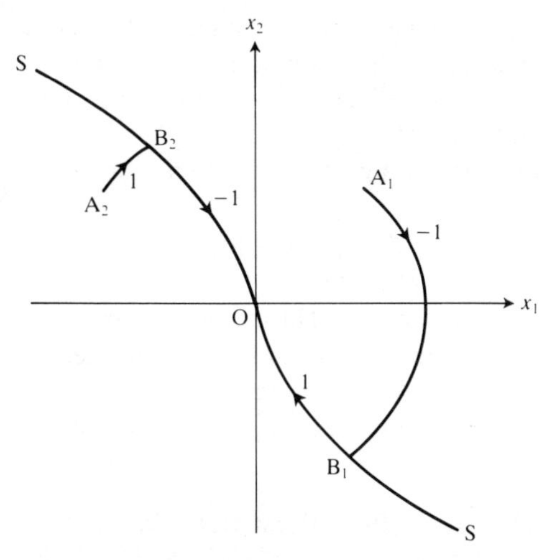

Fig. 6.1. Optimal trajectories from A_1 and A_2. B_1 and B_2 are the points where the control is discontinuous on the switching curve SOS.

curve, the target can be reached with no switching of the control. If it lies to the left of the switching curve, we must begin by applying the control $u_1 = +1$, and switch to $u_1 = -1$ when the trajectory intersects the curve $x_2^2 = -2x_1, x_2 > 0$.

In solving this problem we have not had to make use of condition (iv) of the PMP, although it is a free-time problem. But neither have we determined the constants A and B in the co-state solution. Since the switch occurs when z_2 changes sign, and when $t = t_2$, it follows that

$$B - At_2 = 0. \tag{6.59}$$

The value of H at $t = t_1$ is given by

$$H = -1 + |B - At_1| = 0 \tag{6.60}$$

and these two equations fix the values of A and B, although this information is of no significance.

6.4. The positioning problem with a fuel cost

The cost functions that we have employed so far have been related to the time and to the square of the control variable, separately and together. Another possible cost depends on the absolute value of the control variable, which is often referred to as a fuel cost, since fuel is consumed whatever the direction of the thrust exerted by an engine. Costs of this kind introduce a new feature into the solution and increase the difficulty of the problem.

Consider the positioning problem with a cost function defined by

$$J = \int_0^{t_1} (k + |u_1|)\,\mathrm{d}t, \tag{6.61}$$

where k is a positive constant. Thus we have two components for the cost, one depending on the time and one on the fuel, with a constant k to measure the relative importance of the two components. For simplicity, we shall assume an initial state at rest at a distance X from the origin, so that

$$x_1(0) = X, \quad x_2(0) = 0, \quad x_1(t_1) = 0, \quad x_2(t_1) = 0. \tag{6.62}$$

The state equations are again given by (6.47) and the control belongs to $\mathscr{U}_b$, as before. The Hamiltonian has the form

$$H = -k - |u_1| + z_1 x_2 + z_2 u_1, \tag{6.63}$$

and the co-state equations are unaltered so that we again have

$$z_1 = A, \quad z_2 = B - At. \tag{6.64}$$

The novel feature of this problem appears when we try to determine the supremum of H. We have to choose u_1, which must lie between -1 and $+1$, in order to let the quantity $q \equiv z_2 u_1 - |u_1|$ take its greatest possible value. If u_1 is positive or zero, $q = u_1(z_2 - 1)$ and q has its greatest possible value if $u_1 = 1$ when $z_2 > 1$ and $u_1 = 0$ when $z_2 < 1$. Similarly, if u_1 is negative or zero, $q = u_1(z_2 + 1)$ and we must take $u_1 = -1$ when $z_2 < -1$ and $u_1 = 0$ when $z_2 > -1$. Combining these results, we deduce that the control variable is given by

$$u_1 = \begin{cases} +1 & \text{when } z_2 > +1, \\ 0 & \text{when } -1 < z_2 < +1, \\ -1 & \text{when } z_2 < -1. \end{cases} \tag{6.65}$$

Previously, when the cost function only depended on the terminal time, the control switched between the extreme values of -1 and $+1$. Now there are three possible settings of the control, the extreme values and with the control switched off. With a cost function depending on the magnitude of u_1, it is clearly advantageous to set $u_1 = 0$, if possible, but it is equally clear that, in general, the control must be employed for part of the time. The co-state variable z_2 is a linear function of t; it follows that the settings of the control must follow the sequences $\{-1, 0, +1\}$ or $\{+1, 0, -1\}$. Only part of either of these sequences may be needed in a particular case; for example, it may be possible to reach the target without switching the control at all. But what can be stated definitely is that it is never possible to switch directly from -1 to $+1$ or back again in the optimal solution. Note that the determination of the control in (6.65) leaves its value arbitrary when z_2 is equal to -1 or $+1$. In general, this will only occur at discrete values of t, and the indeterminancy of the control at these instants will not affect the solution. An exception would be if z_2 were a constant. Then, if $z_2 = +1$, u_1 could take any value between 0 and $+1$ and, if $z_2 = -1$, u_1 could take any value between 0 and -1.

With the initial and final conditions prescribed by (6.62), it is clear that we cannot start with $u_1 = 0$, for then we would remain at the initial state. For a similar reason, we cannot end with $u_1 = 0$. Also, an initial positive control would move the state away from the target, and so we are led to the conclusion that we must make the control pass through the complete sequence $\{-1, 0, +1\}$. Hence the initial trajectory is given by

$$x_2 = -t, \quad x_1 = X - \tfrac{1}{2}t^2, \quad x_1 + \tfrac{1}{2}x_2^2 = X, \tag{6.66}$$

and the final section of the trajectory is given by

$$x_2 = t - t_1, \quad x_1 = \tfrac{1}{2}(t_1 - t)^2, \quad x_1 - \tfrac{1}{2}x_2^2 = 0. \tag{6.67}$$

These two trajectories intersect, but we know that they must be separated by a trajectory corresponding to $u_1 = 0$. The state equations (6.47) show that for this part of the solution

$$x_2 = a, \quad x_1 = b + at, \tag{6.68}$$

where a and b are constants. Suppose the switching times are t_2 and t_3, with $0 < t_2 < t_3 < t_1$. Equating the values of x_1 and x_2 at these switching times, to ensure that the state varies continuously, yields the equations

$$\begin{gathered} -t_2 = a, \quad t_3 - t_1 = a, \\ X - \tfrac{1}{2}t_2^2 = b + at_2, \quad \tfrac{1}{2}(t_1 - t_3)^2 = b + at_3, \end{gathered} \tag{6.69}$$

from which we deduce that

$$t_2 + t_3 = t_1, \quad t_2 t_3 = X. \tag{6.70}$$

These equations determine the switching times when we know the value of t_1, but this has still to be determined. To do so, we must consider the co-state variables. We know from (6.65) that the switchings occur when z_2 passes through the values -1 and $+1$. Hence, with z_2 given by (6.64), we see that

$$B - At_2 = -1, \quad B - At_3 = +1, \tag{6.71}$$

so that

$$A = -2/(t_3 - t_2), \quad B = -(t_3 + t_2)/(t_3 - t_2) = -t_1/(t_3 - t_2), \tag{6.72}$$

using (6.70). These results determine the co-state variables, but we still have not found the value of t_1. The one result that we have not yet employed is condition (iv) of the PMP. Since this is a free-time problem, the Hamiltonian is zero along the optimal trajectory. Evaluating it at $t = 0$ when $u_1 = -1$ and $x_2 = 0$ gives

$$-k - 1 - B = 0. \tag{6.73}$$

When this result is combined with the equations in (6.70) and (6.71) we find that

$$\begin{gathered} t_3 - t_2 = pt_1, \quad t_3 = \tfrac{1}{2}pt_1(k+2), \quad t_2 = \tfrac{1}{2}pt_1 k, \\ X = \tfrac{1}{4}k(k+2)p^2 t_1^2, \quad \text{where } p = 1/(k+1). \end{gathered} \tag{6.74}$$

The optimal time is fixed by this last result. The optimal cost, from (6.61), is given by

$$\begin{aligned} J &= kt_1 + t_2 + (t_1 - t_3) \\ &= t_1/p - pt_1 \\ &= 2k^{1/2}(k+2)^{1/2}X^{1/2}. \end{aligned} \tag{6.75}$$

The optimal trajectory is sketched in Fig. 6.2.

If k is large, so that the cost depends mainly on the time taken to reach the target and to a minor extent on the fuel consumed, the interval $t_3 - t_2$ during which the engines are idle is very short and we approach the solution of the positioning problem with time optimization given in §6.3(3). The flat portion of the trajectory in Fig. 6.2 is close to the intersection of the two parabolic curves. If k is small, the saving of fuel is the dominant consideration, and the flat trajectory

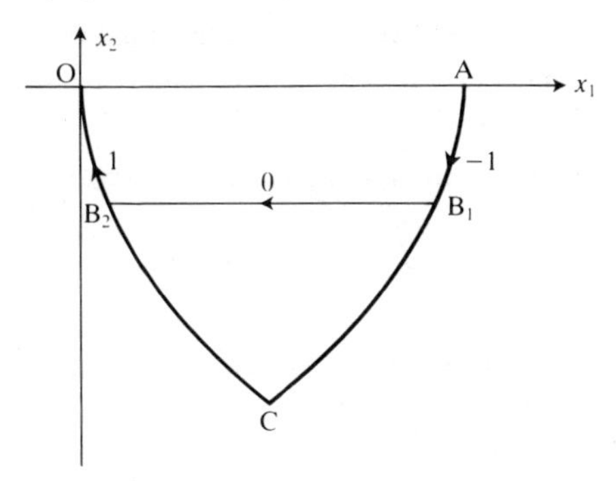

Fig. 6.2. Optimal path from A to O with time and fuel cost. From B_1 to B_2 the control is not applied and no fuel is used. The position of B_1 between A and C depends on the value of k, that is, on the relative importance of time and fuel in the cost.

lies just underneath the x_1 axis. In the limit as k tends to zero, the cost can be made arbitrarily small, but the time to reach the target tends to infinity. If we had attempted to solve the problem with $k = 0$, we would have found that the switching time $t_2 = 0$, so that we would have to begin with the engines switched off. Since this implies that we remain at the initial position for ever, there is no optimal solution, although we could make the cost arbitrarily small. If we switch the engine on for a short time τ with $u_1 = -1$, we reach the point $x_1 = X - \frac{1}{2}\tau^2, x_2 = -\tau$. Then we can switch the engines off and coast along with constant speed τ until we reach $x_1 = \frac{1}{2}\tau^2$. Finally we can use $u_1 = +1$ for a further time τ to reach the origin. The total time taken is equal to $(X - \tau^2)/\tau + 2\tau = X/\tau + \tau$ and the cost is 2τ. Since τ is arbitrary, the cost can be made arbitrarily small, though the time taken is then arbitrarily large.

There are several directions in which the PMP can be extended to a wider class of problems, and we shall discuss these in the next chapter.

Exercises 6

(*Solutions on p. 226*)

6.1 A one-dimensional stable system returns to its equilibrium position in an infinite time, so a control is applied to speed up the restoration of equilibrium. The state equation and cost function are given by

$$\dot{x}_1 = -x_1 + u_1, \quad J = \int_0^{t_1} (k + \tfrac{1}{2}u_1^2)\, dt,$$

where k is a positive constant and $u_1 \in \mathscr{U}_u$. Find the optimal solution for the initial state $x_1^0 = X$, where X is positive. What is the maximum size of the control used in the optimal solution?

6.2 Repeat Exercise 6.1 with $u_1 \in \mathscr{U}_b$ and with the cost function given by

$$J = \int_0^{t_1} (k + |u_1|)\,dt.$$

6.3 Determine the optimal solution for the positioning problem of §6.3(3), with $x_1^0 = X > 0$, $x_2^0 = 0$, and target at the origin, when the cost function is given by

$$J = \int_0^{t_1} (\tfrac{1}{2}k^2 + \tfrac{1}{2}u_1^2)\,dt,$$

and the control belongs to $\mathscr{U}_u$. Find the maximum size of the control used.

Consider the limiting cases when k is very small and very large.

6.4 Repeat Exercise 6.3 with the initial state given by $x_1^0 = 1$, $x_2^0 = 1$.

6.5 Solve the control problem for gout, as specified in §6.3(2), but with the cost function replaced by one that depends only on the state and the time taken to reach the terminal state:

$$J = \int_0^{t_1} \tfrac{1}{2}(k^2 + x_1^2)\,dt.$$

There is a maximum permitted dose and the control satisfies the condition $-2 \le u_1 \le 2$.

6.6 A plague of flies increases exponentially unless spiders are present in sufficient numbers. If the number of flies at time t is $x_1(t)$ and the number of spiders is $x_2(t)$, suitable state equations have the form

$$\dot{x}_1 = x_1 - x_2, \quad \dot{x}_2 = u_1,$$

where u_1 is the controllable rate at which the number of spiders can be increased or reduced. If $|u_1| \le 1$, and $x_1(0) = X$, $x_2(0) = 0$, with $0 < X < 1$, show that a bang–bang control with a single switching is required to eliminate both flies and spiders simultaneously in the shortest possible time. Show that the optimal time t_1 is given by

$$t_1 = -2\ln(1 - X^{1/2}).$$

Comment on the situation when $X \ge 1$.

7 Extensions to the PMP

> A little onward lead thy guiding hand
> To these dark steps, a little further on.
>
> John Milton

The basic form of the PMP can be extended to apply to more general targets than the single point so far assumed. We can also include the effect of a terminal cost, and we can make a simple amendment of the principle to allow non-autonomous problems to be treated.

7.1. General targets

It often happens that we do not need to specify the final state of the system completely. It may be sufficient, for example, that one component only of the state vector must reach a specified value, while the other components are free to take any values consistent with the solution of the state equations. In that case, we lose information about the solution of the state equations, and this lack must be supplied from another source. If not, we would end with an undetermined problem. The extra conditions come from a constraint on the *co-state* vector at the terminal time. These conditions are stated and applied here, but, in accordance with our procedure, the reason for their validity is postponed until Chapter 9, where we give an outline proof of the PMP and its extensions.

Suppose the target is defined as some surface in $\mathscr{R}^n$ of dimension $n - k$. Such a surface can be defined as the intersection of k surfaces each of dimension $n - 1$, with equations of the form

$$g_i(\mathbf{x}) = 0, \quad i = 1, 2, \ldots, k. \tag{7.1}$$

For example, if $n = 3$ and $k = 2$, the two surfaces might be two planes, intersecting in a line, and the target state could be anywhere along this line. Let $\mathbf{x}(t_1) = \mathbf{x}^1$, the final state reached by the system; when the target point is completely known, this provides n conditions on the solution of the state equations. With a partially free target, we know that

$$g_i(\mathbf{x}^1) = 0, \quad i = 1, 2, \ldots, k, \tag{7.2}$$

which provides k conditions. The remaining $n-k$ conditions that are needed to fix the solution are provided by what is known as *the transversality condition.* Suppose that $\mathbf{z}(t_1) = \mathbf{z}^1$, where $\mathbf{z}$ is the co-state vector. The transversality condition states that $\mathbf{z}^1$ is normal to the target set at the (unknown) target point $\mathbf{x}^1$ at the terminal time t_1. This is equivalent to saying that $\mathbf{z}^1$ is linearly dependent on the normals to the surfaces defining the target set, or that

$$\mathbf{z}^1 = \sum_1^k c_i \operatorname{grad} g_i(\mathbf{x}^1) \tag{7.3}$$

for some constants c_i. This provides us with n equations but introduces k extra unknowns, so it gives the $n-k$ conditions we require.

Of course, since (7.3) depends on the position of the target point, the application of the transversality condition may not provide us with an easy route to the solution. If, however, the equations of the surfaces defining the target set are linear in the state variables, the normals to these surfaces point in fixed directions and (7.3) is much easier to apply. To clarify the meaning of the transversality condition, let us consider the case $n = 3$ and all possible values of k. If $k = 3$, suppose the surfaces defining the target set are $x_1 = 0$, $x_2 = 0$, $x_3 = 0$. The target is a fixed point, $\mathbf{x}^1 = \mathbf{0}$, and (7.3) states that $\mathbf{z}^1$ is arbitrary. For $k = 2$, suppose that $g_1(\mathbf{x}) = x_1$ and $g_2(\mathbf{x}) = x_2$. The target is the x_3 axis and (7.3) has the form

$$\mathbf{z}^1 = c_1 \operatorname{grad} g_1 + c_2 \operatorname{grad} g_2 = c_1 \begin{bmatrix} 1 \\ 0 \\ 0 \end{bmatrix} + c_2 \begin{bmatrix} 0 \\ 1 \\ 0 \end{bmatrix}, \tag{7.4}$$

or, equivalently, $\mathbf{z}^1$ lies in the plane $x_3 = 0$. Hence the three conditions to be applied at time t_1 are

$$x_1^1 = 0, \quad x_2^1 = 0, \quad z_3^1 = 0. \tag{7.5}$$

For $k = 1$, suppose that $g_1(\mathbf{x}) = x_1 + x_2$, so that the target set is a plane. Then the transversality condition states that

$$\mathbf{z}^1 = c_1 \operatorname{grad} g_1 = c_1 \begin{bmatrix} 1 \\ 1 \\ 0 \end{bmatrix}, \tag{7.6}$$

so that the three terminal conditions are given by

$$x_1^1 + x_2^1 = 0, \quad z_1^1 - z_2^1 = 0, \quad z_3^1 = 0. \tag{7.7}$$

Finally, suppose that $k = 0$. There are then no conditions on the target point, which is completely free. The condition (7.3) becomes $\mathbf{z}^1 = \mathbf{0}$, and again we have the required three conditions to apply at time t_1. This last case might seem to be of little practical significance since, if the state is free to move to any final position, it could remain where it is initially, and no cost would be incurred in moving it. However, as we shall see in the next section, it is possible that part of the total cost may depend on the terminal state, and the optimal solution may then require the state to move away from its initial position.

In the cases just discussed we see how greater freedom of the state vector is compensated by an increase in the constraint placed on the co-state vector. To show how a problem with a partially fixed target can be solved, consider the following example.

1. *The harmonic oscillator with an energy cost*

Suppose we have a harmonic oscillator with its equilibrium position at the origin, for which the state equations have the form

$$\dot{x} = x_2, \quad \dot{x}_2 = -x_1 + u_1. \tag{7.8}$$

Let us suppose that initially the system is at rest at a distance X from the equilibrium position and that we wish to move to the equilibrium position, without necessarily requiring that we arrive there with zero velocity. The initial and final states are, therefore, given by

$$x_1^0 = X, \quad x_2^0 = 0, \quad x_1^1 = 0, \quad x_2^1 \text{ arbitrary}. \tag{7.9}$$

The transversality condition states that $\mathbf{z}^1$ must be normal to the target set, which is given by $x_1 = 0$, so that

$$\mathbf{z}^1 = c_1 \begin{bmatrix} 1 \\ 0 \end{bmatrix}, \quad \text{or} \quad z_2^1 = 0. \tag{7.10}$$

To complete the specification of the problem, we choose the cost function to be given by

$$J = \int_0^{t_1} \tfrac{1}{2} u_1^2 \, dt, \tag{7.11}$$

so that the supplementary state equation is

$$\dot{x}_0 = \tfrac{1}{2} u_1^2. \tag{7.12}$$

If the terminal time t_1 is free, we can reach the target without applying any control in a quarter-period of the oscillation; the cost is then zero and we cannot do better than that. Suppose, however, that the terminal time t_1 is fixed, and that $t_1 < \frac{1}{2}\pi$. The optimal choice of the control variable is then not obvious.

If we now apply the PMP in the standard manner, we arrive at the co-state equations

$$\dot{z}_1 = z_2, \quad \dot{z}_2 = -z_1 \tag{7.13}$$

and the solution of these equations that satisfies the transversality condition (7.3) can be written as

$$z_1 = A\cos(t_1 - t), \quad z_2 = A\sin(t_1 - t), \tag{7.14}$$

where A is some constant. To maximize the Hamiltonian, defined by

$$H = -\tfrac{1}{2}u_1^2 + z_1 x_2 + z_2(-x_1 + u_1), \tag{7.15}$$

we choose $u_1 = z_2$, and the solution of the state equations (7.8) is easily found to be given by

$$\begin{aligned} x_1 &= \tfrac{1}{2}At\cos(t_1 - t) + X\cos t + B\sin t, \\ x_2 &= \tfrac{1}{2}A\cos(t_1 - t) + \tfrac{1}{2}At\sin(t_1 - t) - X\sin t + B\cos t, \end{aligned} \tag{7.16}$$

which satisfies the initial condition on x_1. The remaining conditions to be satisfied are that $x_2^0 = 0$ and $x_1^1 = 0$. From (7.16), these conditions show that

$$B = -\tfrac{1}{2}A\cos t_1, \quad A = \frac{-2X\cos t_1}{t_1 - \sin t_1 \cos t_1}. \tag{7.17}$$

The optimal cost can be found by substituting the value of z_2 for u_1 in (7.11), which reduces to

$$J = \frac{X^2\cos^2 t_1}{t_1 - \sin t_1 \cos t_1}. \tag{7.18}$$

When $t_1 = \frac{1}{2}\pi$ we see that the optimal cost is indeed zero, as anticipated, and $u_1 = 0$. For $t_1 < \frac{1}{2}\pi$, the optimal cost is positive, and, as $t_1 \to 0$, $J \sim \frac{3}{2}X^2/t_1^3$. A large cost is incurred if we insist on reaching the zero position in a very short time.

We shall meet the harmonic oscillator for a fixed target and with different cost functions later.

7.2. Terminal cost

The cost function, in addition to the part that depends on the time history of the state and control variables, may contain an element that depends on the final state reached by the system. The new form of the cost function is taken to be

$$J = F(\mathbf{x}^1) + \int_0^{t_1} f_0(\mathbf{x}, \mathbf{u})\, \mathrm{d}\tau. \tag{7.19}$$

If the target is fixed, the contribution to the cost from the first term is known, so that the problem is unaffected by this addition. But if the system has some freedom to end at different positions within the target set, this extra term may inhibit the terminal state from going too far. The change in the cost function demands a change in the definition of the extra state variable x_0. We now define it by the additional state equation

$$\dot{x}_0 = f_0 + \sum_1^n \frac{\partial F}{\partial x_i} \dot{x}_i, \quad x_0(0) = x_0^0 = F(\mathbf{x}^0). \tag{7.20}$$

It is clear from these definitions that $x_0(t_1) = x_0^1 = J$.

We can now proceed to apply the PMP in the usual way, although we shall change the notation so that the extended co-state vector is $\hat{\boldsymbol{\zeta}}$ and not $\hat{\mathbf{z}}$. The Hamiltonian $H'(\hat{\mathbf{x}}, \hat{\boldsymbol{\zeta}}, \mathbf{u})$ is now given by

$$H' = \zeta_0 f_0 + \zeta_0 \sum_1^n \frac{\partial F}{\partial x_i} f_i + \sum_1^n \zeta_i f_i. \tag{7.21}$$

The co-state equations and the transversality condition (7.3) are given by

$$\dot{\zeta}_i = -\frac{\partial H'}{\partial x_i}, \quad \boldsymbol{\zeta}^1 = \sum_1^k c_i \operatorname{grad} g_i(\mathbf{x}^1). \tag{7.22}$$

Now let us define an adjusted co-state vector $\hat{\mathbf{z}}$ by the equations

$$z_0 = \zeta_0 = -1, \quad z_i = \zeta_i + \zeta_0 \frac{\partial F}{\partial x_i} \quad \text{for } i = 1, 2, \ldots, n. \tag{7.23}$$

In these new variables the Hamiltonian has the standard form $H = \hat{\mathbf{z}}^{\mathrm{T}}\hat{\mathbf{f}}$, where $H(\hat{\mathbf{x}}, \hat{\mathbf{z}}, \mathbf{u}) \equiv H'(\hat{\mathbf{x}}, \hat{\boldsymbol{\zeta}}, \mathbf{u})$, and the co-state equations are

$$\dot{z}_0 = 0, \quad \dot{z}_i = -\frac{\partial H}{\partial x_i} \quad \text{for } i = 1, 2, \ldots, n, \tag{7.24}$$

while the transversality condition, from (7.22) and (7.23), becomes

$$\mathbf{z}^1 = -\operatorname{grad} F(\mathbf{x}^1) + \sum_1^k c_i \operatorname{grad} g_i(\mathbf{x}^1). \tag{7.25}$$

It follows, therefore, that problems with a terminal cost can be solved without changing the co-state variable. The standard procedure can be applied and the only alteration is that there is an extra term in the transversality condition. Since (7.25) reduces to the usual transversality condition (7.3) when $F = 0$, we can adopt (7.25) as the statement of this condition in all cases.

As an example of a problem with a terminal cost, consider the following case study.

1. *Control of infection*

A patient waiting to undergo an operation has an infection which it is desirable to reduce before the operation is performed. However, the surgeon does not want to delay the operation indefinitely. Left to itself, the infection would die away, but this could take a long time; to speed things up a drug is administered to reduce the infection. There are two conflicting objectives: to reduce the infection, even though this might mean a long delay, and to act quickly, even though the level of the infection is dangerously high. The surgeon has to strike a balance between these two objectives and to decide on the optimal time when the operation should be performed. To put this problem into mathematical form, suppose the level of the infection is measured by a state variable $x_1(t)$, with a positive initial value equal to X. The drug prescribed produces a decrease in the growth rate of the infection equal to $u_1(t)$, and the maximum permitted dose is $u_1 = 1$. The cost function has two components: one is proportional to the time and the other depends on the level of the infection at the terminal time. Suppose we define the cost by

$$J = \tfrac{1}{2}(x_1^1)^2 + \int_0^{t_1} 2\,dt, \tag{7.26}$$

where, as usual, $x_1^1 = x_1(t_1)$. A suitable form for the state equations, that includes both the natural and drug-induced decay of the infection, is

$$\dot{x}_1 = -x_1 - u_1. \tag{7.27}$$

We have now formulated the optimal control problem completely. The target is free, as is the terminal time, and the cost contains an element that depends on the terminal state. We can now apply the PMP, including the transversality condition in the form (7.25), which takes account of the terminal cost.

The Hamiltonian for this problem is given by

$$H = -2 + z_1(-x_1 - u_1), \tag{7.28}$$

and the co-state equation and transversality condition are

$$\dot{z}_1 = z_1, \quad z_1^1 = -x_1^1. \tag{7.29}$$

The solution of the co-state equation is $z_1 = Ae^t$ and we maximize H by taking $u_1 = -\text{sgn}(z_1)$. Since z_1 does not change sign, we can take $u_1 = 1$ and A negative. We could try the other possibility, with the signs reversed, but it leads nowhere; it is clear that there is no point in increasing the infection. Since t_1 is free, the Hamiltonian must be zero along the optimal trajectory, and applying this condition at $t = 0$ gives the equation

$$-2 + A(-X - 1) = 0, \tag{7.30}$$

which determines the value of A. The solution of the state equation (7.27), with $u_1 = 1$ and $x_1^0 = X$, is

$$x_1 = -1 + (X + 1)\,e^{-t} \tag{7.31}$$

and the transversality condition requires that

$$\frac{-2e^{t_1}}{X + 1} = 1 - (X + 1)e^{-t_1}. \tag{7.32}$$

Solving this equation for t_1, we arrive at the values of the terminal time and state in the optimal solution,

$$t_1 = \ln\left[\frac{X + 1}{2}\right], \quad x_1^1 = 1, \tag{7.33}$$

and the optimal cost is

$$J = 2\ln\left[\frac{X + 1}{2}\right] + \tfrac{1}{2}. \tag{7.34}$$

There is, however, a hidden snag in accepting this solution. There is no difficulty if $X > 1$, but if $0 \leq X \leq 1$, $t_1 \leq 0$ and, in formulating the PMP, we required that t_1 be positive. Thus no optimal solution that satisfies the PMP is possible when X is not greater than one. But there must be a best course of action for the surgeon to take, and the way out of this difficulty is to take $t_1 = 0$. The terminal state is then the same as the initial one, $x_1^1 = X$, and the optimal cost is $J = \frac{1}{2}X^2$.

The strategy the surgeon should apply is, therefore, to operate at once if the infection level X is less than or equal to one. If X is greater than one, he should prescribe the maximum dosage of the drug to reduce the infection until its level has been reduced to one.

This special case when the terminal time is equal to the initial time does not occur only when there is a terminal cost. It is also present, though trivially, when the initial state is actually in the target set.

7.3. Non-autonomous problems

For the time-optimal control problems of Part A we are restricted to autonomous problems only, in which the state equations and cost function do not depend explicitly on the time. We have made the same assumption in considering the more general problems covered by the PMP but this restriction can, in fact, be removed. In non-autonomous problems, time can appear in two guises: the state equations depend explicitly on t, and t is also the independent variable involved in the time derivatives. The key idea in the adaptation of the PMP to such problems is to separate these two roles, replacing t where it occurs explicitly by a new dependent variable, while keeping t as the independent variable in the derivatives and in the integral involved in the cost function. This higher-dimensional problem is autonomous.

Suppose the state equations and cost function are given by

$$\dot{\mathbf{x}} = \mathbf{f}(t, \mathbf{x}, \mathbf{u}), \quad J = \int_{t_0}^{t_1} f_0(t, \mathbf{x}, \mathbf{u})\, dt, \tag{7.35}$$

and let $\bar{H}$ be the Hamiltonian for the non-autonomous problem. We introduce the new variable x_{n+1}, for which the state equation and initial value are

$$\dot{x}_{n+1} = 1, \quad x_{n+1}^0 = t_0. \tag{7.36}$$

Note that, because the original problem was not autonomous, the solution does not depend solely on the time interval and we need to specify the initial time t_0, which is not necessarily zero. In the modified problem, the value of t_0 provides a boundary condition for the new variable x_{n+1}.

The extended vectors $\hat{\mathbf{x}}$, $\hat{\mathbf{f}}$, and $\hat{\mathbf{z}}$ now have $n+2$ components, labelled from 0 to $n+1$, and the associated Hamiltonian is defined by

$$H = \hat{\mathbf{z}}^{\mathrm{T}}\hat{\mathbf{f}}. \tag{7.37}$$

The co-state equations are given by

$$\dot{z}_i = -\frac{\partial H}{\partial x_i} = -\sum_{k=0}^{n} z_k \frac{\partial f_k}{\partial x_i} \quad \text{for } i = 1, 2, \ldots, n+1, \tag{7.38}$$

and $\dot{z}_0 = 0$ with $z_0 = -1$, as before. The last component of the co-state vector appears only in the last of the co-state equations (7.38), so that we can write down its value in the form

$$z_{n+1} = -\int_{t_1}^{t} \sum_{0}^{n} z_k \frac{\partial f_k(\tau, \mathbf{x}, \mathbf{u})}{\partial \tau} \mathrm{d}\tau + z_{n+1}^1. \tag{7.39}$$

The target in the $(n+1)$-dimensional space is given by x_i^1 for $i = 1, 2, \ldots, n$, which is fixed, and $x_{n+1}^1 = t_1$, which may be either fixed or free. If t_1 is free, we must apply the transversality condition. The target set is a line in $\mathscr{R}^{n+1}$, so that this condition tells us that $z_{n+1}^1 = 0$. Since $f_{n+1} = 1$, the modified Hamiltonian has the form

$$H = \sum_{0}^{n} z_i f_i + z_{n+1} = \bar{H} + z_{n+1}. \tag{7.40}$$

If we apply the condition that $H = 0$ along the optimal trajectory when $t = t_1$, the transversality condition shows that $\bar{H} = 0$. Now suppose that t_1 is fixed. The PMP tells us that H is constant along the optimal trajectory, but also the value of z_{n+1}^1 is unspecified. We can therefore insist that $H = 0$ and fix the value of z_{n+1}^1 accordingly. This consideration of the non-autonomous problem helps to explain the difference we met before between the fixed- and free-time problems. In this more general setting, we see that $H = 0$ in all cases. Note, however, that in a non-autonomous problem it is not necessarily true that $\bar{H}$ is constant along the optimal trajectory since z_{n+1} may vary with time.

Here is a simple example of a non-autonomous problem.

1. *Filling a reservoir*

Water from a reservoir is being drawn off at an ever-increasing rate. In addition to compensating for this water loss, it is required to raise the level of the water in the reservoir above its initial height by pumping in fresh water. The cost of this operation per unit time is proportional to the square of the pumping rate, and we wish to adjust this rate to minimize the cost. The required level must be reached in a specified time; alternatively, the time to fill the reservoir may be of no concern. A set of model equations for this system, with all the parameters eliminated by suitable scalings, can be stated in the following manner.

Consider the state equation

$$\dot{x}_1 = -v + u_1, \quad x_1(0) = 0, \quad x_1(t_1) = 1, \tag{7.41}$$

together with the cost function

$$J = \int_0^{t_1} \tfrac{1}{2}u_1^2 \, \mathrm{d}t. \tag{7.42}$$

The height of the water above the initial level is measured by x_1, and we want to reach the height 1 in time t_1. The pumping rate is given by $u_1(t)$, and $v(t)$ is the rate at which water is drawn off. Suppose we put $v = t$ and consider both the fixed-time problem $t_1 = 1$ and the free-time problem.

We can write this system as an autonomous one with three variables, so that

$$\begin{bmatrix} \dot{x}_0 \\ \dot{x}_1 \\ \dot{x}_2 \end{bmatrix} = \begin{bmatrix} \tfrac{1}{2}u_1^2 \\ -x_2 + u_1 \\ 1 \end{bmatrix}, \quad \hat{\mathbf{x}}^0 = \begin{bmatrix} 0 \\ 0 \\ 0 \end{bmatrix}, \quad \hat{\mathbf{x}}^1 = \begin{bmatrix} J \\ 1 \\ t_1 \end{bmatrix}. \tag{7.43}$$

The Hamiltonian for this problem has the form

$$H = z_0 \tfrac{1}{2}u_1^2 + z_1(-x_2 + u_1) + z_2 \tag{7.44}$$

and the co-state equations are

$$\dot{z}_0 = 0, \quad \dot{z}_1 = 0, \quad \dot{z}_2 = z_1. \tag{7.45}$$

The solutions of these equations are

$$z_0 = -1, \quad z_1 = A, \quad z_2 = At + B, \tag{7.46}$$

where A and B are constants. The supremum of H is attained when $u_1 = z_1 = A$, and the solutions of the state equations are

$$x_0 = \tfrac{1}{2}A^2 t, \quad x_1 = -\tfrac{1}{2}t^2 + At, \quad x_2 = t. \tag{7.47}$$

The Hamiltonian, evaluated along the optimal trajectory, is given by

$$H = -\tfrac{1}{2}A^2 + A(-t + A) + B + At = B + \tfrac{1}{2}A^2. \qquad (7.48)$$

The condition that H be zero and the known terminal value of x_1 yield the equations

$$B + \tfrac{1}{2}A^2 = 0, \quad -\tfrac{1}{2}t_1^2 + At_1 = 1. \qquad (7.49)$$

We can now consider the fixed and free terminal times. With $t_1 = 1$, $A = \frac{3}{2}$ and $J = \frac{9}{8}$. The condition $H = 0$ then serves to fix $B = -\frac{9}{8}$ and so to determine the value of z_2, but this is of no particular interest. With t_1 free, we employ the transversality condition, which shows that $z_2^1 = 0$. We now have three equations for the three unknowns A, B, and t_1, and their values are given by

$$A = \frac{2\sqrt{2}}{\sqrt{3}}, \quad B = -\frac{4}{3}, \quad t_1 = \frac{\sqrt{2}}{\sqrt{3}}; \qquad (7.50)$$

the optimal cost has the value

$$J = \frac{4\sqrt{2}}{3\sqrt{3}} = 1.089. \qquad (7.51)$$

This optimal cost is less than the value of 1.125 when the terminal time was fixed, as of course it must be, but the difference is not large.

As well as allowing the state equations and the cost function to depend explicitly on the time, we can also deal in a similar way with a target set which is time dependent. The defining functions $g_i(\mathbf{x})$ that were used in §7.1 can be modified to include their dependence on t as well as on $\mathbf{x}$. With the additional variable x_{n+1} is introduced, these functions depend on the extended state vector, and the way in which general targets are dealt with, as explained in §7.1, is unaltered. The following example shows how this can be done.

2. *A one-dimensional pursuit problem*

A stationary police car is passed by a motorist who maintains a constant and excessive speed. The police car sets off in pursuit, and the driver wants to get into a position alongside the other car, so that they are then moving at the same speed. This manoeuvre is to be performed at minimum cost as measured by a combination of the time and the energy expended. The state equations for the police car can be taken as those used in the positioning problem of §6.3(3), so that

$$\dot{x}_1 = x_2, \quad \dot{x}_2 = u_1, \quad \dot{x}_3 = 1, \tag{7.52}$$

where we have introduced a third state variable x_3 equal to the time. The initial state is given by $\mathbf{x}^0 = \mathbf{0}$, and the target set is given by the state vector for the car being pursued, which can be taken as

$$x_1 = t = x_3, \quad x_2 = 1, \tag{7.53}$$

where we have chosen the constant speed of the car as the unit of velocity. The transversality condition (7.25) states that, at the time of contact with the target,

$$\begin{bmatrix} z_1^1 \\ z_2^1 \\ z_3^1 \end{bmatrix} = c_1 \begin{bmatrix} 1 \\ 0 \\ -1 \end{bmatrix} + c_2 \begin{bmatrix} 0 \\ 1 \\ 0 \end{bmatrix}, \tag{7.54}$$

where the three-dimensional vector $\mathbf{z}$ is the co-state variable. It follows that the three conditions to be satisfied at the terminal time t_1 are

$$x_1^1 = t_1, \quad x_2^1 = 1, \quad z_1^1 + z_3^1 = 0. \tag{7.55}$$

Let us take a cost function for this problem in the form

$$J = \int_0^{t_1} (2 + \tfrac{1}{2}u_1^2)\,\mathrm{d}t, \tag{7.56}$$

and the Hamiltonian then becomes

$$H = -(2 + \tfrac{1}{2}u_1^2) + z_1 x_2 + z_2 u_1 + z_3. \tag{7.57}$$

We must take $u_1 = z_2$ to let H attain its supremum. The co-state equations are

$$\dot{z}_1 = 0, \quad \dot{z}_2 = -z_1, \quad \dot{z}_3 = 0, \tag{7.58}$$

and the solutions of these equations, subject to the third condition in (7.55), can be written as

$$z_1 = A, \quad z_2 = B - At, \quad z_3 = -A, \tag{7.59}$$

where A and B are constants. It is now easy to write down the solutions of the state equations (7.52), using $u_1 = z_2$, and when we satisfy the first two conditions in (7.55) we arrive at the equations

$$t_1 = \tfrac{1}{2}Bt_1^2 - \tfrac{1}{6}At_1^3, \quad 1 = Bt_1 - \tfrac{1}{2}At_1^2. \tag{7.60}$$

There are three unknowns, A, B, and t_1, so we need one more equation. This is provided by the condition that H be zero along the optimal trajectory. When the values obtained so far are substituted into the expression for H in (7.57), we obtain the result

$$-2 + \tfrac{1}{2}B^2 - A = 0. \tag{7.61}$$

The solution of these equations is

$$A = 6, \quad B = 4, \quad t_1 = 1. \tag{7.62}$$

The optimal control and the optimal trajectory are given by

$$u_1 = 4 - 6t, \quad x_1 = 2t^2 - t^3, \quad x_2 = 4t - 3t^2, \tag{7.63}$$

and the optimal cost is

$$J = \int_0^1 [2 + \tfrac{1}{2}(4 - 6t)^2]\,\mathrm{d}t = 4. \tag{7.64}$$

The pursuit ends when $x_1 = 1, x_2 = 1$. If, for $t > 1$, the control u_1 is switched off, the two cars will move along the road side by side.

In this chapter we have generalized the basic PMP to enable us to apply it to an extended class of optimal control problems. The examples chosen have been restricted, for simplicity, to systems with a small number of degrees of freedom. It is not often possible to obtain specific results for problems with many components, but there is one class for which we can make considerable progress. This topic is sufficiently important to merit a chapter to itself.

Exercises 7

(*Solutions on p. 229*)

7.1 Use the PMP to solve the problem of Exercise 1.1.

7.2 Find the time-optimal solution for the positioning problem, with state equations $\dot{x}_1 = x_2$, $\dot{x}_2 = u_1$, and with $|u_1| \leq 1$, for the general initial state given by $x_1 = X_1, x_2 = X_2$. Consider the case in which the target is given by $x_2 = 0$.

Repeat this example for the target set $x_1 = -1$ with $X_1 > -1$.

7.3 A three-dimensional control problem has state equations given by

$$\dot{x}_1 = x_2, \quad \dot{x}_2 = x_3, \quad \dot{x}_3 = u_1,$$

and the cost function is

$$J = \int_0^{t_1} (k + \tfrac{1}{2}u_1^2)\, dt \quad (k \text{ positive}).$$

If the initial state is (X_1, X_2, X_3) with $X_3 > 0$, and the target is the plane $x_3 = 0$, show that the optimal cost is given by

$$J = (2k)^{1/2} X_3$$

and determine the terminal time and the final state of the system.

7.4 Consider the state equations given in Exercise 7.3, but with the cost function defined by

$$J = \int_0^{t_1} (k + |u_1|)\, dt.$$

With the same initial and final conditions as in Exercise 7.3, determine the optimal solution.

7.5 The state equations for the harmonic oscillator are given by

$$\dot{x}_1 = x_2, \quad \dot{x}_2 = -x_1 + u_1.$$

Suppose that the control u_1 lies between $+1$ and -1, and that the cost function is given by

$$J = \int_0^{t_1} (1 + |u_1|)\, dt.$$

If the initial state is given by $x_1^0 = 0$, $x_2^0 = X > 0$, and the target state lies on the line $x_2^1 = 0$, show that the optimal cost when $X < 1$ has the value $2 \tan^{-1} X$.

Find the optimal cost when $X > 1$ and show that the optimal control is not uniquely determined when $X = 1$.

7.6 A target point P is moving with uniform speed round a circular track of radius r, so that its position at time t is given by $x_1 = r \cos t$, $x_2 = r \sin t$. A pursuer Q is initially at the origin and can move with velocity u_1 in the x_1 direction and u_2 in the x_2 direction, both components having a maximum magnitude equal to one. Q wishes to reach P in as short a time as possible. Show that, if $r = \pi/\sqrt{8}$, the optimal control is given by $u_1 = 1$, $u_2 = 1$, and that the optimal time is $t_1 = \frac{1}{4}\pi$.

Find the optimal solution when $r = \frac{1}{2}\pi$.

7.7 In the standard positioning problem, the applied force can take any permissible value and can switch instantaneously. Suppose instead that the force can only increase or decrease at a specified maximum rate, but there is no limit on the magnitude of the force. Show that state equations that represent this system can have the form

$$\dot{x}_1 = x_2, \quad \dot{x}_2 = x_3, \quad \dot{x}_3 = u_1,$$

with $|u_1| \leq 1$. If it is required to meet the target $x_1 = 0$, $x_2 = 0$ in minimum time, show that the control is bang–bang with at most one switch. Find the optimal solution when the initial state is given by $x_1 = X > 0$, $x_2 = 0$, $x_3 = 0$.

If in the final state x_3 is also to be reduced to zero, show that two switches may be needed and find the optimal solution with the same initial conditions.

7.8 A miser stores a fraction u of the interest on his investments in a box under his bed. He re-invests what remains after paying tax at a rate λ on the unstored proceeds. After T years he dies and estate duty at a rate μ is charged on the total of his investments and what he has stored. If the interest rate is k and the value of his investments is $x(t)$, show that

$$\dot{x} = k(1 - \lambda)ux$$

and that the terminal value $e(T)$ of his estate and the total tax paid $f(T)$ are given by

$$e(T) = x(T) + \int_0^T (1 - u)kx \, \mathrm{d}t, \quad f(T) = \mu e(T) + \int_0^T \lambda ukx \, \mathrm{d}t.$$

The terminal time T is supposed known and the initial capital $x(0) = X$. Use the PMP for a problem with a free target and a terminal cost to determine the optimal choice of the control $u(t)$ if it is desired to minimize the total tax paid.

Find also the optimal strategy if it is desired to maximize the final value of the estate.

7.9 The force acting on an object that needs to be repositioned increases linearly with the time, and a control force of bounded magnitude can be applied to it. The state equations are

$$\dot{x}_1 = x_2, \quad \dot{x}_2 = t + u_1, \quad |u_1| \leq 1,$$

with the initial state given by $x_1^0 = X > 0$, $x_2^0 = 0$. The final state $x_1^1 = 0$, $x_2^1 = 0$ is to be reached in the shortest possible time t_1. Show that t_1 is a root of the equation

$$t_1^2(12 + 4t_1 - 3t_1^2) = 48X.$$

Show that an optimal solution is only possible when $X \leq \frac{2}{3}$, and describe the outcome when $X > \frac{2}{3}$.

7.10 A harmonic oscillator is subjected to a resonating periodic force and the state equations with an additional controlling force have the form

$$\dot{x}_1 = x_2, \quad \dot{x}_2 = -x_1 + 2\cos t + u_1.$$

The initial state is the origin and, with $u_1 = 0$, the solution is $x_1 = t \sin t$. It is desired to control the system to the original state after the fixed time t_1 with minimum cost J, defined by

$$J = \int_0^{t_1} u_1^2 \, dt.$$

It is obvious that the control $u_1 = -2\cos t$ is a successful control for any value of t_1. Show that it is also the optimal control, with an optimal cost given by

$$J = 2t_1 + \sin 2t_1.$$

8 Linear state equations with quadratic costs

A poet does not work by square or line.

William Cowper

The combination of quadratic cost functions and linear state equations provides an important class of problems for which the application of the PMP leads to a set of linear equations. We have met particular examples of this kind already, but here we try to find general solutions. We deal first with the simplest case, before considering n-dimensional problems.

8.1. The one-dimensional problem

In the one-dimensional problem of the kind to be studied in this chapter, there is one state variable x_1 and one control variable u_1. We suppose that the initial time t_0 and state x_1^0 are given, that the terminal state x_1^1 is free, and that the terminal time t_1 is either fixed or free. The state equation is linear in both the state and control variables and has the form

$$\dot{x}_1 = ax_1 + bu_1, \tag{8.1}$$

where a and b are constants. The cost contains a terminal part, as well as a part that depends quadratically on the state and control variables. We define it by

$$J(t_0, t_1) = \tfrac{1}{2}f(x_1^1)^2 + \frac{1}{2}\int_{t_0}^{t_1} (qx_1^2 + ru_1^2)\,\mathrm{d}t, \tag{8.2}$$

where f and q are non-negative constants and r is a positive constant.

The Hamiltonian for this problem has the form

$$H = -\tfrac{1}{2}(qx_1^2 + ru_1^2) + z_1(ax_1 + bu_1), \tag{8.3}$$

and the supremum is attained when

$$-ru_1 + bz_1 = 0. \tag{8.4}$$

Hence the state and co-state equations are given by

$$\dot{x}_1 = ax_1 + b^2r^{-1}z_1, \tag{8.5}$$

$$\dot{z}_1 = qx_1 - az_1, \tag{8.6}$$

and these equations are to be solved subject to the given initial condition and the transversality condition related to the terminal cost, as described in §7.2. These conditions are that

$$x_1(t_0) = x_1^0, \quad z_1^1 + fx_1^1 = 0, \tag{8.7}$$

where $x_1(t_1) = x_1^1$ and $z_1(t_1) = z_1^1$, as usual. We are faced, therefore, with the task of solving a pair of first-order linear differential equations with constant coefficients. This poses no difficulties, except that we do not have an initial-value problem and so cannot be sure that we shall be able to satisfy the conditions (8.7). If no such solution exists, then we can deduce that there is no optimal solution of the control problem we have posed. In order to test whether or not this is so, we can write down the general solution of the equations (8.5) and (8.6) in the form

$$\begin{aligned} x_1 &= A_1 \exp[\lambda(t - t_0)] + A_2 \exp[-\lambda(t - t_0)], \\ z_1 &= B_1 \exp[\lambda(t - t_0)] + B_2 \exp[-\lambda(t - t_0)], \end{aligned} \tag{8.8}$$

where

$$\begin{aligned} b^2r^{-1}B_1 = (\lambda - a)A_1, \quad b^2r^{-1}B_2 = (\lambda + a)A_2, \\ \lambda = (a^2 + qb^2r^{-1})^{1/2}. \end{aligned} \tag{8.9}$$

Applying the conditions (8.7) and solving for A_1 and A_2, we obtain

$$\begin{aligned} A_1 &= x_1^0 \exp(-\lambda T)(\lambda + a - fb^2r^{-1})/D, \\ A_2 &= x_1^0 \exp(\lambda T)(\lambda - a + fb^2r^{-1})/D, \end{aligned} \tag{8.10}$$

where $T = t_1 - t_0$ and

$$D = 2\lambda \cosh \lambda T + 2(fb^2r^{-1} - a) \sinh \lambda T. \tag{8.11}$$

Since $\lambda > a$, and f and r are positive, $D > 0$, and we have a solution of the boundary-value problem for all choices of the constants.

The optimal cost can be found directly by integration, or more simply as follows. From the equations (8.5) and (8.6) we see that

$$\frac{\mathrm{d}}{\mathrm{d}t}(z_1x_1) = z_1\dot{x}_1 + x_1\dot{z}_1 = b^2r^{-1}z_1^2 + qx_1^2 = ru_1^2 + qx_1^2, \quad (8.12)$$

so that

$$\begin{aligned} J &= \tfrac{1}{2}f(x_1^1)^2 + \frac{1}{2}\int_{t_0}^{t_1} \frac{\mathrm{d}}{\mathrm{d}t}(z_1x_1)\,\mathrm{d}t \\ &= \tfrac{1}{2}f(x_1^1)^2 + \tfrac{1}{2}z_1^1x_1^1 - \tfrac{1}{2}z_1^0x_1^0 \\ &= -\tfrac{1}{2}z_1^0x_1^0, \end{aligned} \quad (8.13)$$

where we have made use of the transversality condition. The value of z_1^0 can be found from (8.8) and (8.9) and, after some simplification, we obtain the result

$$J(t_0, t_1) = \tfrac{1}{2}(x_1^0)^2 f \frac{\lambda + (qf^{-1} + a)\tanh \lambda T}{\lambda + (fb^2r^{-1} - a)\tanh \lambda T}. \quad (8.14)$$

This completes the solution of the fixed-time problem. It is easy to see that $J(t_0, t_1)$ is a monotonic function of t_1. If we define S by

$$S = \frac{\lambda + qf^{-1} + a}{\lambda + fb^2r^{-1} - a}, \quad (8.15)$$

the fraction in (8.14) varies between the values 1, when $t_1 = t_0$, and S, when $t_1 \to \infty$. It follows that there is no solution of the free-time optimal control problem with the terminal time finite and greater than the initial time. If $S > 1$, the minimum value of $J(t_0, t_1)$ as a function of t_1 is the initial cost, equal to $\tfrac{1}{2}(x_1^0)^2 f$, when $t_1 = t_0$. If $S < 1$, then $J(t_0, t_1)$ is always greater than $\tfrac{1}{2}(x_1^0)^2 fS$ and approaches this value as $t_1 \to \infty$ without ever reaching it. If $S = 1$, the optimal cost is equal to the initial cost and can be achieved at any terminal time by a suitable choice of the control variable. The condition $S > 1$ is equivalent to

$$q > f(fb^2r^{-1} - 2a), \quad (8.16)$$

which follows from the definition of S.

Instead of finding the minimum of $J(t_0, t_1)$ as a function of t_1, it is, of course, possible to arrive at these results for the free-time problem by applying the usual condition on the Hamiltonian (see Exercise 8.1).

There is another method for treating problems of the sort we have been considering, which we shall refer to as the *K-method*. Consider the fixed-time problem again. The transversality condition in (8.7) suggests that it might be possible to obtain a solution by introducing a new variable $K(t, t_1)$ defined by

$$z_1 = -Kx_1. \tag{8.17}$$

The transversality condition then becomes

$$K(t_1, t_1) = f, \tag{8.18}$$

and the optimal cost, from (8.13), is given by

$$J(t_0, t_1) = \tfrac{1}{2}(x_1^0)^2 K(t_0, t_1). \tag{8.19}$$

Since $\dot{z}_1 = -\dot{K}x_1 - K\dot{x}_1$, the state and co-state equations (8.5) and (8.6) show that K satisfies the differential equation

$$\dot{K} = b^2 r^{-1} K^2 - 2aK - q. \tag{8.20}$$

If this first-order equation is solved subject to the 'initial' condition in (8.18), $K(t_0, t_1)$ can be found, and then the optimal cost follows at once from (8.19), for any initial state.

We can represent the manner in which K varies by sketching the graph of $\dot{K}$ against K as shown in Fig. 8.1. Define K_c by

$$K_c = \frac{ar}{b^2}\left[1 + \left(1 + \frac{b^2 q}{a^2 r}\right)^{1/2}\right], \tag{8.21}$$

so that $\dot{K} = 0$ when $K = K_c$. Then we can distinguish two cases. First suppose that $0 < f < K_c$. Then $K(t_0, t_1)$ lies between f and K_c and the path in the phase plane is from A at $t = t_0$ to B at $t = t_1$. The optimal cost is given in terms of $K(t_0, t_1)$, which is the value of K at A. We see from the figure that it increases as the time interval T, which is equal to $t_1 - t_0$, increases and its smallest value as a function of T is $\frac{1}{2}f(x_1^0)^2$ with $T = 0$. Secondly, suppose that $f > K_c$. Then the path is from C at t_0 to D at t_1, and the optimal cost is a decreasing function of T, tending to the value $\frac{1}{2}K_c(x_1^0)^2$ as $T \to \infty$. These features of the solution are, of course, the same as those found using the first method.

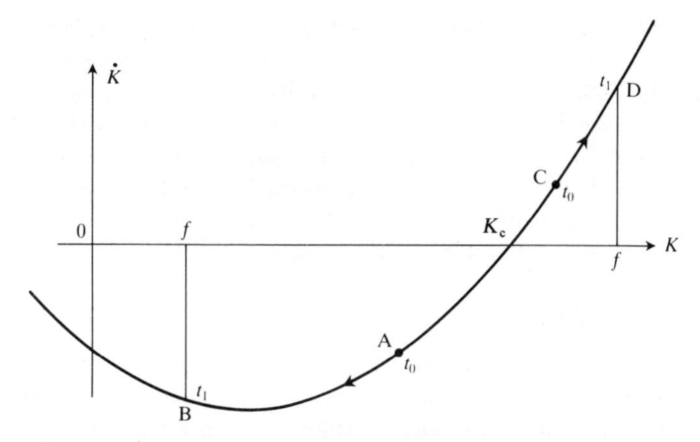

Fig. 8.1. $\dot{K}$ as a function of K from (8.20) for two different values of f.

The solution of (8.20) can easily be found by separating the variables. It has the form

$$K(t, t_1) = arb^{-2} + \lambda rb^{-2} \tanh[\lambda(t_1 - t) + c], \qquad (8.22)$$

where λ has the value given in (8.9), and, since $K(t_1, t_1) = f$, the constant c is given by

$$\lambda rb^{-2} \tanh c = f - arb^{-2}. \qquad (8.23)$$

The value of $K(t_0, t_1)$ can then be determined and when it is substituted into (8.19) we recover the expression (8.14) for the optimal cost.

The two methods we have used for the one-dimensional problem with a free target can be applied to problems in which the target is fixed; see Exercises 8.2 and 8.3. Using the first method, we have to solve two first-order linear equations, while the K-method required the solution of a single, but nonlinear, equation. Since linear equations are generally much easier to solve than nonlinear ones, the second method would not seem to have any advantage over the first. There is, however, another consideration which adds to the value of the K-method. The linear equations of the first method have to be solved with boundary conditions, not initial ones. That is, we have a certain number of conditions to be satisfied at the initial time t_0 and the remainder at the terminal time t_1. In the particular problem we have solved, we were able to find a solution satisfying these conditions, because the denominator D in (8.10) did not vanish. In general, we cannot be certain that a solution will exist, as was explained in §2.1. The K-method, however, resulted in a first-order equation, and the only condition that had to be satisfied was at the terminal time t_1. Hence we have an initial-value problem, and we know that a solution will exist, subject to the requirements of Picard's theorem (see §2.1). This holds also for n-dimensional problems, so that when we apply the K-method we can be certain that a solution will exist, even though it may not be easy to find.

8.2. The n-dimensional problem

The general form of the linear-state problems with quadratic costs that we are studying has an n-dimensional state vector $\mathbf{x}$ and an m-dimensional control vector $\mathbf{u}$. The state equation is linear in both the state and control variables and has the general form

$$\dot{\mathbf{x}} = \mathbf{A}\mathbf{x} + \mathbf{B}\mathbf{u}, \tag{8.24}$$

where $\mathbf{A}$ is an $n \times n$ matrix and $\mathbf{B}$ is an $n \times m$ matrix. The elements in these matrices are, in general, time dependent. The cost consists of a terminal part and a part that depends quadratically on the state and control variables. We define the cost as the value of $J(t_0, t_1)$, where the cost function $J(t, t_1)$ is given by

$$J(t, t_1) = \tfrac{1}{2}\mathbf{x}^{1\mathrm{T}}\mathbf{F}\mathbf{x}^1 + \frac{1}{2}\int_t^{t_1} (\mathbf{x}^{\mathrm{T}}\mathbf{Q}\mathbf{x} + \mathbf{u}^{\mathrm{T}}\mathbf{R}\mathbf{u})\, \mathrm{d}\tau. \tag{8.25}$$

In this expression, $\mathbf{F}$ is a constant $n \times n$ matrix, $\mathbf{Q}$ is a time-dependent $n \times n$ matrix, and $\mathbf{R}$ is a time-dependent $m \times m$ matrix. The quadratic forms with matrices $\mathbf{F}$ and $\mathbf{Q}$ are positive semi-definite and that with matrix $\mathbf{R}$ is positive definite (see §2.5). Since $\mathbf{R}$ is non-singular, $\mathbf{R}^{-1}$ exists. The initial and final times are t_0 and t_1, and $\mathbf{x}^0$ and $\mathbf{x}^1$ are the corresponding states.

We introduce the extra state variables x_0 and x_{n+1}, since this is a non-autonomous problem, satisfying the state equations

$$\dot{x}_0 = \mathbf{x}^{\mathrm{T}}\mathbf{F}\dot{\mathbf{x}} + \tfrac{1}{2}\mathbf{x}^{\mathrm{T}}\mathbf{Q}(t)\mathbf{x} + \tfrac{1}{2}\mathbf{u}^{\mathrm{T}}\mathbf{R}(t)\mathbf{u}, \quad x_0(t_0) = \tfrac{1}{2}\,\mathbf{x}^{0\mathrm{T}}\mathbf{F}\mathbf{x}^0, \tag{8.26}$$

and

$$\dot{x}_{n+1} = 1, \quad x_{n+1}(t_0) = t_0. \tag{8.27}$$

The Hamiltonian for this problem has the form

$$H = \tfrac{1}{2}z_0(\mathbf{x}^{\mathrm{T}}\mathbf{Q}\mathbf{x} + \mathbf{u}^{\mathrm{T}}\mathbf{R}\mathbf{u}) + \mathbf{z}^{\mathrm{T}}(\mathbf{A}\mathbf{x} + \mathbf{B}\mathbf{u}) + z_{n+1}. \tag{8.28}$$

The co-state equations are

$$\dot{z}_0 = 0, \quad \dot{\mathbf{z}} = -z_0\mathbf{Q}\mathbf{x} - \mathbf{A}^{\mathrm{T}}\mathbf{z}, \quad \dot{z}_{n+1} = -\frac{\partial H}{\partial x_{n+1}} = -\frac{\partial H}{\partial t}, \tag{8.29}$$

and the supremum of H is attained when

$$-\mathbf{R}\mathbf{u} + \mathbf{B}^{\mathrm{T}}\mathbf{z} = \mathbf{0}, \tag{8.30}$$

which comes from setting $\partial H/\partial \mathbf{u}$ to zero with $z_0 = -1$. Hence

$$\mathbf{u} = \mathbf{R}^{-1}\mathbf{B}^{\mathrm{T}}\mathbf{z} \tag{8.31}$$

and the state and co-state equations have the forms

$$\dot{\mathbf{x}} = \mathbf{A}\mathbf{x} + \mathbf{B}\mathbf{R}^{-1}\mathbf{B}^{\mathrm{T}}\mathbf{z}, \quad \dot{\mathbf{z}} = \mathbf{Q}\mathbf{x} - \mathbf{A}^{\mathrm{T}}\mathbf{z}. \tag{8.32}$$

The cost function can be found as follows. From (8.32) we see that

$$\begin{aligned}\frac{\mathrm{d}}{\mathrm{d}t}(\mathbf{x}^{\mathrm{T}}\mathbf{z}) &= \mathbf{x}^{\mathrm{T}}(\mathbf{Q}\mathbf{x} - \mathbf{A}^{\mathrm{T}}\mathbf{z}) + (\mathbf{x}^{\mathrm{T}}\mathbf{A}^{\mathrm{T}} + \mathbf{z}^{\mathrm{T}}\mathbf{B}\mathbf{R}^{-1}\mathbf{B}^{\mathrm{T}})\mathbf{z} \\ &= \mathbf{x}^{\mathrm{T}}\mathbf{Q}\mathbf{x} + \mathbf{z}^{\mathrm{T}}\mathbf{B}\mathbf{u} \\ &= \mathbf{x}^{\mathrm{T}}\mathbf{Q}\mathbf{x} + \mathbf{u}^{\mathrm{T}}\mathbf{R}\mathbf{u},\end{aligned} \tag{8.33}$$

where we have used the transpose of the first equation in (8.32) and the fact that $\mathbf{R}$ is symmetric. It follows from (8.25) that

$$J(t, t_1) = \tfrac{1}{2}\mathbf{x}^{1\mathrm{T}}\mathbf{F}\mathbf{x}^1 + \tfrac{1}{2}\mathbf{x}^{1\mathrm{T}}\mathbf{z}^1 - \tfrac{1}{2}\mathbf{x}^{\mathrm{T}}\mathbf{z}, \tag{8.34}$$

and the optimal cost is given by

$$J(t_0, t_1) = \tfrac{1}{2}\mathbf{x}^{1\mathrm{T}}\mathbf{F}\mathbf{x}^1 + \tfrac{1}{2}\mathbf{x}^{1\mathrm{T}}\mathbf{z}^1 - \tfrac{1}{2}\mathbf{x}^{0\mathrm{T}}\mathbf{z}^0, \tag{8.35}$$

where $\mathbf{z}^0$ and $\mathbf{z}^1$ are the values of the co-state variables at t_0 and t_1, respectively.

If we have a fixed-time problem, the condition that H should vanish along the optimal trajectory only serves to determine z_{n+1}. For a free-time problem, the transversality condition associated with the unspecified value of x^1_{n+1} (see §7.3) shows that $z^1_{n+1} = 0$, and the condition on H becomes

$$\tfrac{1}{2}\mathbf{z}^{\mathrm{T}}\mathbf{B}\mathbf{R}^{-1}\mathbf{B}^{\mathrm{T}}\mathbf{z} - \tfrac{1}{2}\mathbf{x}^{\mathrm{T}}\mathbf{Q}\mathbf{x} + \mathbf{z}^{\mathrm{T}}\mathbf{A}\mathbf{x} = 0, \tag{8.36}$$

when the value of $\mathbf{u}$ from (8.31) is used.

The state and co-state equations (8.32) together form a linear homogeneous system of dimension $2n$. Hence we can write down the general solution once we have determined the fundamental matrix (see §2.2). We need $2n$ conditions to complete the solution and the prescribed value of $\mathbf{x}^0$ provides n of them. For a fixed target the remaining conditions come from the known value of $\mathbf{x}^1$. For a target set that is not a single point we have to satisfy the transversality conditions of §7.1 if there is no terminal cost, and the modified form derived in §7.2 if $\mathbf{F} \neq \mathbf{0}$. In general, this will not give linear relations between the components of $\mathbf{x}^1$ and $\mathbf{z}^1$; only when the target set is linear (a line, plane, or hyperplane) will we retain the linearity of the problem. For a free target, the transversality condition becomes $\mathbf{z}^1 + \mathbf{F}\mathbf{x}^1 = \mathbf{0}$. In all cases, therefore, we end with the required number of conditions, n at the initial time t_0 and n at the terminal time t_1.

This is as far as we can go in the determination of the solution in general. Although we can, in theory, obtain the fundamental matrix and hence write down the solution of the general initial-value problem, there is no guarantee that we shall be able to choose the unknown initial value of $\mathbf{z}^0$ so as to satisfy the conditions at t_1. The extension of the K-method introduced for one-dimensional problems in §8.1 to n dimensions provides an alternative approach in certain cases, and then we can be sure that a solution exists.

8.3. The *K*-method

The K-method can be applied to the general problem defined in §8.2, restricted to the case of a free target and a fixed terminal time. The condition to be satisfied at t_1 is then

$$\mathbf{z}^1 = -\mathbf{F}\mathbf{x}^1, \tag{8.37}$$

and this suggests that we write, as in (8.17),

$$\mathbf{z} = -\mathbf{K}(t, t_1)\mathbf{x}. \tag{8.38}$$

The transversality condition (8.37) thus becomes

$$\mathbf{K}(t_1, t_1) = \mathbf{F}. \tag{8.39}$$

If we differentiate (8.38) with respect to t and substitute the values of $\dot{\mathbf{x}}$ and $\dot{\mathbf{z}}$ from (8.32) we obtain the equation

$$\mathbf{Q}\mathbf{x} - \mathbf{A}^{\mathrm{T}}\mathbf{K}\mathbf{x} = -\dot{\mathbf{K}}\mathbf{x} - \mathbf{K}[\mathbf{A}\mathbf{x} + \mathbf{B}\mathbf{R}^{-1}\mathbf{B}^{\mathrm{T}}(-\mathbf{K}\mathbf{x})]. \tag{8.40}$$

Since this equation must hold for all values of $\mathbf{x}$ reached from arbitrary initial values $\mathbf{x}^0$, we deduce that $\mathbf{K}$ must satisfy the differential equation

$$\dot{\mathbf{K}} = \mathbf{K}\mathbf{B}\mathbf{R}^{-1}\mathbf{B}^{\mathrm{T}}\mathbf{K} - \mathbf{K}\mathbf{A} - \mathbf{A}^{\mathrm{T}}\mathbf{K} - \mathbf{Q}. \tag{8.41}$$

This is a first-order system of differential equations with an initial value given by (8.39), so a solution exists. Once $\mathbf{K}$ has been found, the optimal trajectory can be determined by solving the initial-value problem

$$\dot{\mathbf{x}} = \mathbf{A}\mathbf{x} - \mathbf{B}\mathbf{R}^{-1}\mathbf{B}^{\mathrm{T}}\mathbf{K}\mathbf{x}, \quad \mathbf{x}(t_0) = \mathbf{x}^0. \tag{8.42}$$

The cost function $J(t, t_1)$ is given by (8.34) and (8.38), so that

$$J(t, t_1) = \tfrac{1}{2}\mathbf{x}^{\mathrm{T}}(t)\mathbf{K}(t, t_1)\mathbf{x}(t), \tag{8.43}$$

and the optimal cost is therefore

$$J(t_0, t_1) = \tfrac{1}{2}\mathbf{x}^{0\mathrm{T}}\mathbf{K}(t_0, t_1)\mathbf{x}^0. \tag{8.44}$$

Some general properties of $\mathbf{K}$ can now be deduced. The definition (8.25) of the cost function ensures that it is positive definite and hence, from (8.43), we see that, for $t < t_1$, $\mathbf{K}(t, t_1)$ is positive definite. Also, $\mathbf{K}(t_1, t_1)$ is positive definite, except when $\mathbf{F} = \mathbf{0}$. Another property of $\mathbf{K}$ is that it is a symmetric matrix. For, if we take the transpose of (8.41) and use the fact that both $\mathbf{Q}$ and $\mathbf{R}$ are symmetric, we obtain the differential equation for $\mathbf{K}^{\mathrm{T}}$, namely

$$\dot{\mathbf{K}}^{\mathrm{T}} = \mathbf{K}^{\mathrm{T}}\mathbf{B}\mathbf{R}^{-1}\mathbf{B}^{\mathrm{T}}\mathbf{K}^{\mathrm{T}} - \mathbf{A}^{\mathrm{T}}\mathbf{K}^{\mathrm{T}} - \mathbf{K}^{\mathrm{T}}\mathbf{A} - \mathbf{Q}. \tag{8.45}$$

In addition, $\mathbf{K}^{\mathrm{T}}(t_1, t_1) = \mathbf{F}^{\mathrm{T}} = \mathbf{F}$, so that $\mathbf{K}^{\mathrm{T}}$ satisfies the same equation as $\mathbf{K}$ and the same initial condition. By the uniqueness of the solution of the initial-value problem, it follows that $\mathbf{K}^{\mathrm{T}}(t, t_1) = \mathbf{K}(t, t_1)$.

The equation (8.41) satisfied by $\mathbf{K}$ is a nonlinear equation and a solution is only guaranteed within some time interval, so t_0 must be sufficiently close to t_1. The following special case is one for which we can be certain of the existence of a solution for all time intervals. We consider an autonomous problem with no terminal cost. Hence, $\mathbf{F} = \mathbf{0}$ and $\mathbf{A}$, $\mathbf{B}$, $\mathbf{R}$, and $\mathbf{Q}$ are constant matrices, and $J(t, t_1)$ is an increasing function of the time difference $T = t_1 - t$ only. We want to establish that $\mathbf{K}$, which is also a function of T only, is bounded for all finite times and tends to a finite limit as $T \to \infty$.

In this problem the target point is free. Let us suppose that the initial state $\mathbf{x}^0$ can be steered to $\mathbf{0}$ by some acceptable control $\mathbf{v}$ in the time interval $(t_0, t_0 + \tau)$. In the terminology of §3.1, the initial state is controllable to the origin, and the conditions given there for this to be true are satisfied. (This is the only place in Part B where we have to make use of results established in Part A.) Since we can ensure that the state remains at the origin by subsequently applying a zero control, we can move the state from $\mathbf{x}^0$ to $\mathbf{0}$ in any time $T > \tau$ and no additional cost is involved during the interval from $t_0 + \tau$ to $t_0 + T$. If we denote the total cost in this manoeuvre by J_0, we have established that there is a possible finite cost J_0 for all $T > \tau$ (recall that $\mathbf{0}$ is a possible target state for the problem under discussion) and hence the optimal cost $J(t_0, t_0 + T) \leq J_0$. But the optimal cost is an increasing function of T, and we have shown that it is bounded above, so that $J(t_0, t_0 + T)$ has a limit as $T \to \infty$, say J_∞. Since the optimal cost is given by (8.44), it follows that $\mathbf{K}(t_0, t_1)$ tends to a limit, $\mathbf{K}_c$ say, as $t_1 - t_0 \to \infty$. From the differential equation satisfied by $\mathbf{K}$, it follows that

$$\mathbf{K}_c\mathbf{B}\mathbf{R}^{-1}\mathbf{B}^{\mathrm{T}}\mathbf{K}_c - \mathbf{A}^{\mathrm{T}}\mathbf{K}_c - \mathbf{K}_c\mathbf{A} - \mathbf{Q} = \mathbf{0}. \tag{8.46}$$

To illustrate these results, let us consider a variant of the positioning problem.

1. *The positioning problem with state and energy costs*

The state equations for this problem are

$$\dot{x}_1 = x_2, \quad \dot{x}_2 = u_1, \tag{8.47}$$

with $\mathbf{x}^{0\mathrm{T}} = (X_1, X_2)$ and a free target state. The cost is given by

$$J = \frac{1}{2}\int_{t_0}^{t_1} (4x_1^2 + u_1^2)\,\mathrm{d}t. \tag{8.48}$$

The matrices defining this system are thus given by

$$\mathbf{A} = \begin{bmatrix} 0 & 1 \\ 0 & 0 \end{bmatrix}, \quad \mathbf{B} = \begin{bmatrix} 0 \\ 1 \end{bmatrix}, \quad \mathbf{R} = [1], \quad \mathbf{Q} = \begin{bmatrix} 4 & 0 \\ 0 & 0 \end{bmatrix}. \tag{8.49}$$

We can write the matrix $\mathbf{K}$ in the form

$$\mathbf{K} = \begin{bmatrix} k_{11} & k_{12} \\ k_{12} & k_{22} \end{bmatrix}, \tag{8.50}$$

where we have used the symmetry of $\mathbf{K}$ to reduce the number of unknown elements to three. When we substitute (8.49) and (8.50) into (8.41), we find, after some algebra, that the differential equations satisfied by the elements of $\mathbf{K}$ are

$$\dot{k}_{11} = k_{12}^2 - 4, \quad \dot{k}_{12} = k_{12}k_{22} - k_{11}, \quad \dot{k}_{22} = k_{22}^2 - 2k_{12}, \tag{8.51}$$

and the initial conditions are

$$k_{11} = 0, \quad k_{12} = 0, \quad k_{22} = 0, \quad \text{when } t = t_1. \tag{8.52}$$

To find $\mathbf{K}_c$ we set $\dot{\mathbf{K}} = \mathbf{0}$. The first equation in (8.51) then gives $k_{12} = \pm 2$, but the third equation shows that k_{12} must be positive, so that

$$k_{12} = 2, \quad k_{22} = \pm 2, \quad k_{11} = \pm 4. \tag{8.53}$$

Hence we have two solutions for $\mathbf{K}_c$,

$$\mathbf{K}_c = \begin{bmatrix} 4 & 2 \\ 2 & 2 \end{bmatrix} \quad \text{and } \mathbf{K}_c = \begin{bmatrix} -4 & 2 \\ 2 & -2 \end{bmatrix}, \tag{8.54}$$

but only the first of these satisfies the requirement that $\mathbf{K}_c$ be positive definite. It follows that, in the limit as $t_1 - t_0 \to \infty$, the optimal cost tends to the value

$$J_\infty = \tfrac{1}{2}[X_1\ X_2]\begin{bmatrix} 4 & 2 \\ 2 & 2 \end{bmatrix}\begin{bmatrix} X_1 \\ X_2 \end{bmatrix} = 2X_1^2 + 2X_1X_2 + X_2^2. \tag{8.55}$$

Of course, since J is an increasing function of the time difference the minimum optimal time for the free-time problem is when $t_1 = t_0$, for the cost is then zero. This result is similar to that found in §8.1 when we were examining the one-dimensional problem, and there was no terminal cost.

It is natural to ask if the K-method can be extended to problems with a fixed target. In general, the answer is no, but there is one exception which we examine in the next section.

8.4. The *K*-method for a fixed target at the origin

The success of the K-method for free-target problems hinged on the form of the transversality condition, which suggested the mapping from $\mathbf{x}$ to $\mathbf{z}$. The key feature of the method is that the value of the mapping matrix was known at t_1. For a fixed target, this is no longer the case, except when $\mathbf{x}^1 = \mathbf{0}$. The previous substitution is not appropriate, but instead we use it in inverted form. We write

$$\mathbf{x} = -\mathbf{L}(t, t_1)\mathbf{z}, \quad \mathbf{L}(t_1, t_1) = \mathbf{0}. \tag{8.56}$$

We can then proceed in the same way as before, and we find that $\mathbf{L}$ satisfies the differential equation

$$\dot{\mathbf{L}} = \mathbf{LQL} + \mathbf{LA}^{\mathrm{T}} + \mathbf{AL} - \mathbf{BR}^{-1}\mathbf{B}^{\mathrm{T}}. \tag{8.57}$$

If $\mathbf{L}$ is any solution of this equation and $\mathbf{K}$ any solution of (8.41), it can be verified that $\mathbf{LK} = \mathbf{C}$, where $\mathbf{C}$ is a constant matrix. The difference, however, is that in solving for $\mathbf{L}$ we must satisfy the initial condition in (8.56), so that $\mathbf{L}$ is singular at $t = t_1$, whereas, in solving for $\mathbf{K}$ when $\mathbf{F} = \mathbf{0}$, it is $\mathbf{K}$ that is singular at $t = t_1$. The optimal cost is given by (8.44), so that

$$J(t_0, t_1) = \tfrac{1}{2}\mathbf{x}^{0\mathrm{T}}\mathbf{L}^{-1}(t_0, t_1)\mathbf{x}^0. \tag{8.58}$$

As before, it can be shown that $\mathbf{L}$ is symmetric and positive definite. For the autonomous case, $\mathbf{L}$ is an increasing function of T, the time difference $t_1 - t_0$. If $\mathbf{L} \to \mathbf{L}_c$ as $T \to \infty$ and if

$$J_\infty = \tfrac{1}{2}\mathbf{x}^{0\mathrm{T}}\mathbf{L}_c^{-1}\mathbf{x}^0, \tag{8.59}$$

J_∞ is the infimum of the optimal costs for all T, but is not attained for any finite time interval.

If we solve the problem discussed in §8.3(1) with a fixed target at the origin instead of the free target postulated there, the equations for the components of $\mathbf{L}$ have the form

$$\dot{l}_{11} = 4l_{11}^2 + 2l_{12}, \quad \dot{l}_{12} = 4l_{11}l_{12} + l_{22}, \quad \dot{l}_{22} = 4l_{12}^2 - 1. \tag{8.60}$$

To find $\mathbf{L}_c$ we set the time derivatives equal to zero and use the fact that the matrix is positive definite. Hence, we find that

$$\mathbf{L}_c = \begin{bmatrix} \frac{1}{2} & -\frac{1}{2} \\ -\frac{1}{2} & 1 \end{bmatrix}, \quad \mathbf{L}_c^{-1} = \begin{bmatrix} 4 & 2 \\ 2 & 2 \end{bmatrix}, \tag{8.61}$$

and the cost infimum is given by

$$J_\infty = 2X_1^2 + 2X_1X_2 + X_2^2, \tag{8.62}$$

where $\mathbf{x}^{0\mathrm{T}} = (X_1, X_2)$. This is the same value as we found for the free-target problem (8.55). If we take T arbitrarily large, we can steer the initial state to the origin with a cost that is greater than, but arbitrarily close to, J_∞.

To conclude this chapter we return to the control of diabetes, which was formulated in §1.5 as an optimal control problem with linear state equations and a quadratic cost.

8.5. Diabetes mellitus (continued from Chapter 1)

The simplified model problem for the control of diabetes was formulated in §1.5. When the variables are redefined to bring them into line with our standard form, the statement of the problem is as follows. The state equations are

$$\dot{x}_1 = -c_1 x_1 - c_2 x_2, \quad \dot{x}_2 = -c_3 x_2 + c_4 x_1 + u_1, \tag{8.63}$$

where x_1 and x_2 are the excess levels of glucose and hormone, respectively. The coefficients are positive constants and the presence of the disease results in the coefficient c_4 being greatly reduced from its value in a healthy individual. We can consider here the extreme case and set $c_4 = 0$. The rate at which insulin is injected is measured by u_1 and we are assuming in this model that it is administered continuously. One possible choice of initial and final states is given by

$$x_1(0) = X_1, \quad x_2(0) = X_2, \quad x_1(t_1) = 0, \quad x_2(t_1) = 0, \tag{8.64}$$

and the terminal time t_1 is fixed. The cost is given by

$$J = \frac{1}{2}\int_0^{t_1} (x_1^2 + k^2 u_1^2)\, dt, \tag{8.65}$$

which contains a part that depends on the excess amount of glucose present and a part that depends on the amount of drug injected, with a weighting factor to measure the relative importance of these two components of the total cost.

The problem just formulated is an autonomous two-dimensional problem with linear state equations and a single control variable. The cost is quadratic, and the target and terminal time are both fixed. Solving this problem in the usual way, we find that the state and co-state equations have the form

$$\begin{bmatrix} \dot{x}_1 \\ \dot{x}_2 \\ \dot{z}_1 \\ \dot{z}_2 \end{bmatrix} = \begin{bmatrix} -c_1 & -c_2 & 0 & 0 \\ 0 & -c_3 & 0 & k^{-2} \\ 1 & 0 & c_1 & 0 \\ 0 & 0 & c_2 & c_3 \end{bmatrix} \begin{bmatrix} x_1 \\ x_2 \\ z_1 \\ z_2 \end{bmatrix}. \tag{8.66}$$

The eigenvalues of this 4×4 matrix are the four values of λ satisfying the equation

$$\lambda^2 = \tfrac{1}{2}(c_1^2 + c_3^2) \pm [\tfrac{1}{4}(c_1^2 - c_3^2)^2 - k^{-2}c_2^2]^{1/2}, \tag{8.67}$$

and all the eigenvalues are real when

$$c_2 \leq \tfrac{1}{2}\, k|c_1^2 - c_3^2|; \tag{8.68}$$

otherwise, the eigenvalues are all complex. The state and co-state variables can be found from the corresponding eigenvectors and from them we can construct the solution for any set of initial and final conditions and for all values of the coefficients c_1, c_2, c_3, and k.

For illustrative purposes, let us consider a special case, choosing the coefficients c_1 and c_3 to be zero, c_2 equal to one, and the initial value of x_2 to be zero. Then the eigenvalues have the four values $(\pm 1 \pm \mathrm{i})p$, where $2kp^2 = 1$. Solving for the state variables, we find that the solution satisfying the initial conditions has the form

$$\begin{aligned} x_1 &= X_1 \cosh pt \cos pt + A(\cosh pt \sin pt - \sinh pt \cos pt) \\ &\quad + B \sinh pt \sin pt, \\ x_2 &= -2pA \sinh pt \sin pt - p(B - X_1)\cosh pt \sin pt \\ &\quad - p(B + X_1)\sinh pt \cos pt, \end{aligned} \tag{8.69}$$

where A and B are constants. The corresponding co-state variables are given by

$$\begin{aligned} z_1 &= -2pkA \cosh pt \cos pt + pk(B + X_1)\cosh pt \sin pt \\ &\quad + pk(X_1 + B)\sinh pt \cos pt, \\ z_2 &= kX_1 \sinh pt \sin pt - kA(\cosh pt \sin pt + \sinh pt \cos pt) \\ &\quad - kB \cosh pt \cos pt. \end{aligned} \tag{8.70}$$

The optimal cost, from (8.35), is given by

$$J = -\tfrac{1}{2}(x_1^0 z_1^0 + x_2^0 z_2^0) = pkAX_1, \tag{8.71}$$

and the value of A can be found from (8.69) when the final conditions $x_1(t_1) = x_2(t_1) = 0$ are applied. The optimal cost is

$$J = (\tfrac{1}{2}k)^{1/2} X_1^2 \frac{\sinh 2pt_1 + \sin 2pt_1}{\sinh 2pt_1 - \sin 2pt_1}. \tag{8.72}$$

The choice of constants used in this example relates to a system in which the insulin level is entirely controlled by the injected amount, without loss. Similarly, there is no decrease in the glucose level, apart from that provided by the insulin. Since the initial and final levels of insulin are both zero, the control must take negative values for some of the time. This is not a realistic model, and it would be better to drop the terminal condition on the insulin level, that is, to remove the condition $x_2^1 = 0$. If we do this, the target set must lie on the line $x_1 = 0$, and the condition on the final value of x_2 is replaced by a transversality condition, as in §7.1. The co-state vector at the terminal time must be normal to the target set, so that $z_2^1 = 0$. The optimal cost is again given by (8.71) since, in both cases, $x_2^1 z_2^1 = 0$, but A must now be found from the first equation in (8.69) and the second equation in (8.70) at $t = t_1$. Solving these equations, we find that the optimal cost is given by

$$J = (\tfrac{1}{2}k)^{1/2} X_1^2 \frac{\cosh 2pt_1 + \cos 2pt_1}{\sinh 2pt_1 - \sin 2pt_1}, \tag{8.73}$$

and the final insulin level by

$$x_2^1 = 4pX_1 \frac{\sinh 2pt_1 \sin pt_1}{\sinh 2pt_1 - \sin 2pt_1}. \tag{8.74}$$

The control variable has the value

$$u_1 = \frac{X_1}{k} \times \frac{\cosh pt \sin p\tau - \sinh pt \cos p\tau + \cos pt \sinh p\tau - \sin pt \cosh p\tau}{\sinh 2pt_1 - \sin 2pt_1}, \tag{8.75}$$

where $\tau = 2t_1 - t$. In order that insulin be added to the system and not removed, we must ensure that u_1 is non-negative for $0 \leq t \leq t_1$. Examination of the value for u_1 given by (8.75) shows that this is true when $pt_1 \leq \frac{1}{2}\pi$. If this condition is not satisfied, it would be necessary to reformulate the problem, restricting the set of controls and thereby making the application of the PMP more difficult.

Now that we have formulated a variety of optimal control problems and used the PMP to solve them, we must turn our attention to the question of the validity of this principle. An outline proof is given in the next chapter.

Exercises 8

(*Solutions on p. 233*)

8.1 Discuss the solution of the free-time one-dimensional problem defined in §8.1 using the condition on the Hamiltonian instead of considering the minimum of the cost for the fixed-time problem as done in the text.

8.2 Solve the one-dimensional problem of §8.1 with a fixed target at the origin. The terminal cost is no longer relevant. Consider both the fixed-time and the free-time cases.

8.3 Apply the K-method to the fixed-target problem of Exercise 8.2 by writing

$$x_1 = -L(t, t_1)z_1, \quad \text{with } L(t_1, t_1) = 0.$$

Find the differential equation satisfied by L and solve it. What is the optimal cost for the fixed-time problem?

8.4 Solve the one-dimensional optimal control problem with state equation and cost function given by

$$\dot{x}_1 = u_1, \quad J = \int_0^{t_1} \tfrac{1}{2}(x_1^2 + u_1^2 + c^2)\,dt,$$

where the cost is the sum of a quadratic part and a part proportional to the time. The initial state is $x_1(0) = X$, the terminal time t_1 is free, and the target is $x_1(t_1) = 0$.

Solve also the free-target problem with a cost function J_1, defined by

$$J_1 = J + \tfrac{1}{2}(x_1^1)^2,$$

so that a terminal cost is included.

8.5 The state equations for the harmonic oscillator have the form

$$\dot{x}_1 = x_2, \quad \dot{x}_2 = -x_1 + u_1.$$

If the terminal time t_1 is fixed, the initial state is given by $x_1^0 = X_1$, $x_2^0 = X_2$, and the target is $x_1^1 = 0$, $x_2^1 = 0$, find the optimal solution when the cost is given by

$$J = \int_0^{t_1} \tfrac{1}{2}u_1^2\,dt.$$

8.6 Consider the problem specified in Exercise 8.5, except that the cost is now defined by

$$J = \int_0^{t_1} (\tfrac{1}{2}u_1^2 + \tfrac{1}{2}x_1^2)\,dt.$$

Find the limit of the optimal cost as $t_1 \to \infty$.

8.7 Consider the state equations

$$\dot{x}_1 = x_2 + u_1, \quad \dot{x}_2 = x_1 + u_2,$$

and the cost function

$$J = \int_0^T \tfrac{1}{2}(u_1^2 - 2u_1u_2 + u_2^2)\,\mathrm{d}t.$$

The initial and final conditions are $x_1(0) = X_1, x_2(0) = X_2$, and $x_1(T) = x_2(T) = 0$. Show that the supremum of H provides only one equation for the optimal values of the two controls, which cannot be determined uniquely. Show that the optimal cost is determined uniquely. (This example demonstrates what may happen when the matrix $\mathbf{R}$ of the cost function (8.25) is only semi-definite and the equation (8.30) does not have a unique solution.)

9 Proof of the Pontryagin maximum principle

> We must never assume that which is incapable of proof.
>
> George Henry Lewes

We now have to justify the PMP, which we stated in Chapter 6 and extended in Chapter 7. First we show that the PMP applied to linear autonomous time-optimal control problems is identical to the maximum principle TOP established in Part A. Then we outline the proof of the PMP in its basic form as defined in Chapter 6.

9.1. Time-optimal autonomous linear problems

In Part A we proved the existence of a solution of the time-optimal control problem and established a maximum principle TOP which we could use to construct the optimal solution for a variety of problems. In Chapter 6 we introduced the Pontryagin maximum principle which can be applied to a much wider class of problems. We have not, as yet, shown that the PMP provides identical solutions to those obtained using the results of Chapter 4. This identification of the two methods provides corroborative evidence for the validity of the PMP and also leads to a result that will be important when we proceed to the proof of the PMP. It is for this reason that the application of the PMP to time-optimal problems has been delayed until this chapter; otherwise, it could have formed part of Chapter 6.

The state equations for the general linear autonomous problem have the form

$$\dot{\mathbf{x}} = \mathbf{A}\mathbf{x} + \mathbf{B}\mathbf{u}, \tag{9.1}$$

where $\mathbf{A}$ and $\mathbf{B}$ are constant matrices of orders $n \times n$ and $n \times m$, respectively, and the m control variables u_i are piecewise-continuous functions that satisfy the conditions $-1 \leq u_i \leq 1$. The cost is equal to the time taken to reach the target, so

$$J = \int_0^{t_1} 1 \, \mathrm{d}t. \tag{9.2}$$

To apply the PMP to this problem we define the Hamiltonian by

$$H = -1 + \mathbf{z}^{\mathrm{T}}(\mathbf{Ax} + \mathbf{Bu}), \tag{9.3}$$

where $\mathbf{z}$ is the co-state variable. Remembering that the components of $\mathbf{u}$ are bounded by ± 1, we see that the supremum of H is achieved when we choose

$$u_i = \operatorname{sgn}[\mathbf{z}^{\mathrm{T}}\mathbf{B}]_i, \tag{9.4}$$

from which is follows that the controls only take their extreme values, that is, they are bang–bang controls. The co-state equations are

$$\dot{\mathbf{z}} = -\frac{\partial H}{\partial \mathbf{x}} = -\mathbf{A}^{\mathrm{T}}\mathbf{z}, \tag{9.5}$$

and the solution that satisfies the condition $\mathbf{z}(t_1) = \mathbf{z}^1$ is given by

$$\mathbf{z}(t) = \exp[-\mathbf{A}^{\mathrm{T}}(t - t_1)]\mathbf{z}^1. \tag{9.6}$$

If we define a vector $\mathbf{h}$ by the equation

$$\mathbf{h} = \exp(\mathbf{A}^{\mathrm{T}}t_1)\mathbf{z}^1, \tag{9.7}$$

we can use (9.6) and (9.7) to rewrite the condition (9.4) on the optimal controls in the form

$$u_i = \operatorname{sgn}[\mathbf{h}^{\mathrm{T}} \exp(-\mathbf{A}t)\mathbf{B}]_i, \tag{9.8}$$

which is identical to the key result (4.7) found in Chapter 4. Hence the PMP provides an alternative route to the conclusions reached by the methods derived in that chapter. From the definition of the vector $\mathbf{h}$ in §4.2, if follows that $\mathbf{z}^1$ is in the direction of the normal to the reachable set at time t_1.

The co-state equations (9.5) are adjoint to the homogeneous state equations

$$\dot{\mathbf{x}} = \mathbf{Ax}, \tag{9.9}$$

in the sense of §2.4, and it was proved in that section that the scalar product of the solutions of the state and co-state equations is invariant, that is,

$$\mathbf{z}^{\mathrm{T}}\mathbf{x} = \mathbf{z}^{1\mathrm{T}}\mathbf{x}^1, \tag{9.10}$$

where $\mathbf{x}^1 = \mathbf{x}(t_1)$. In the present application of this result, $\mathbf{A}$ is a constant matrix and the fundamental matrix for (9.9) is $\exp(\mathbf{A}t)$. But the invariant property (9.10) holds also for non-constant matrices $\mathbf{A}$, as shown in §2.4, and it is this more general result that will be used in the course of the argument to establish the PMP.

9.2. Statement of the theorem

We consider the optimal control of an autonomous system with given initial state at time $t = 0$ and a fixed target to be reached at time t_1, where t_1 is either fixed or free. The control variables u_i are piecewise-continuous functions of t and may or may not belong to a bounded set. We shall write $u_i \in \mathscr{U}$, where $\mathscr{U}$ may be $\mathscr{U}_{\mathrm{u}}$ or $\mathscr{U}_{\mathrm{b}}$. Controls in $\mathscr{U}_{\mathrm{u}}$ can take any finite value; those in $\mathscr{U}_{\mathrm{b}}$ must satisfy the conditions $-1 \leq u_i \leq 1$. For each successful solution of the state equations using an admissible control we can calculate the value of a cost function, and the optimal solution is one that minimizes the cost. Let $x_0(t)$ satisfy the equation and initial condition

$$\dot{x}_0 = f_0(\mathbf{x}, \mathbf{u}), \quad x_0(0) = x_0^0 = 0, \tag{9.11}$$

so that the cost is given by

$$J = x_0^1 = \int_0^{t_1} f_0(\mathbf{x}, \mathbf{u})\, \mathrm{d}t. \tag{9.12}$$

The state equations and associated boundary conditions have the form

$$\dot{\mathbf{x}} = \mathbf{f}(\mathbf{x}, \mathbf{u}), \quad \mathbf{x}(0) = \mathbf{x}^0, \quad \mathbf{x}(t_1) = \mathbf{x}^1, \tag{9.13}$$

where $\mathbf{x}^0$ and $\mathbf{x}^1$ are both known. The co-state vector is denoted by $\mathbf{z}$, and the extended vectors $\hat{\mathbf{x}}$, $\hat{\mathbf{z}}$, and $\hat{\mathbf{f}}$ are vectors of dimension $n + 1$ defined by

$$\hat{\mathbf{x}} = \begin{bmatrix} x^0 \\ \mathbf{x} \end{bmatrix}, \quad \hat{\mathbf{z}} = \begin{bmatrix} z^0 \\ \mathbf{z} \end{bmatrix}, \quad \hat{\mathbf{f}} = \begin{bmatrix} f^0 \\ \mathbf{f} \end{bmatrix}, \tag{9.14}$$

so that the combined state and cost variables satisfy the equation

$$\dot{\hat{\mathbf{x}}} = \hat{\mathbf{f}}(\mathbf{x}, \mathbf{u}). \tag{9.15}$$

The Hamiltonian H is defined by

$$H(\hat{\mathbf{x}}, \hat{\mathbf{z}}, \mathbf{u}) = \hat{\mathbf{z}}^{\mathrm{T}}\,\hat{\mathbf{f}}(\mathbf{x}, \mathbf{u}), \tag{9.16}$$

and the state and co-state equations are given by

$$\dot{\hat{\mathbf{x}}} = \frac{\partial H}{\partial \hat{\mathbf{z}}}, \quad \dot{\hat{\mathbf{z}}} = -\frac{\partial H}{\partial \hat{\mathbf{x}}}. \tag{9.17}$$

The PMP is a necessary condition, so we assume that an optimal solution exists. For simplicity of notation let $\mathbf{u}$ be the optimal control and $\hat{\mathbf{x}}$ the associated trajectory, while other controls and trajectories are denoted by $\mathbf{v}$ and $\hat{\mathbf{y}}$, respectively. Then the PMP states that there is a solution of the state and co-state equations (9.17) with $z_0 = -1$ and that the optimal control is determined by the maximum principle

$$H(\hat{\mathbf{x}}, \hat{\mathbf{z}}, \mathbf{u}) = \sup_{\mathbf{v} \in \mathscr{U}} H(\hat{\mathbf{x}}, \hat{\mathbf{z}}, \mathbf{v}). \tag{9.18}$$

Moreover, the Hamiltonian, evaluated along the optimal trajectory, is constant so that

$$H[\hat{\mathbf{x}}(t), \hat{\mathbf{z}}(t), \mathbf{u}(t)] = \text{constant}, \tag{9.19}$$

and this constant is zero if t_1 is free. This is the theorem that we wish to prove. The proof outlined here is a simplified version of that given in more precise form by Macki and Strauss (see the Bibliography).

The essence of the argument is that, if we perturb the control away from its optimal value, then, even if we can still reach the target, we incur a greater cost than when using the optimal control. The main difficulty in establishing the result is that the optimal control is not necessarily continuous. Indeed, in many of the examples we have found that the optimal control is bang–bang. A perturbation of such a solution may involve a change in the time at which the switch occurs, leading to a large change in the value of the control then. To avoid the difficulty of a perturbed control that is not always close to the unperturbed control, we consider perturbations that may be large in magnitude but are confined to small time intervals. The net effect on the trajectory of such perturbations will then be small.

9.3. The perturbation cone

We consider here the effect on the trajectory of making a small variation of the kind described in §9.2. We pick a time τ between 0 and t_1 and, for $\tau - \varepsilon < t < \tau$, change the control from the optimal $\mathbf{u}$ to another admissible control $\mathbf{v}$. The small number ε is positive, and we can ensure that the control $\mathbf{u}$ does not have a discontinuity in this time interval (we do not insist that $\mathbf{u}$ be continuous at $t = \tau$). Then we can trace two trajectories, starting from the same point $\hat{\mathbf{x}}(\tau - \varepsilon)$ and ending at time τ at $\hat{\mathbf{x}}(\tau)$ when we use the control $\mathbf{u}$ and at $\hat{\mathbf{y}}(\tau)$ when we use the control $\mathbf{v}$. To the first order in ε,

$$\hat{\mathbf{x}}(\tau) = \hat{\mathbf{x}}(\tau - \varepsilon) + \int_{\tau-\varepsilon}^{\tau} \hat{\mathbf{f}}(\mathbf{x}, \mathbf{u})\,\mathrm{d}t, \quad \hat{\mathbf{y}}(\tau) = \hat{\mathbf{x}}(\tau - \varepsilon) + \int_{\tau-\varepsilon}^{\tau} \hat{\mathbf{f}}(\mathbf{x}, \mathbf{v})\,\mathrm{d}t. \tag{9.20}$$

We denote the difference between these two values by $\hat{\boldsymbol{\xi}}(\tau)$ and it follows that, to the same order in ε,

$$\hat{\boldsymbol{\xi}}(\tau) = \varepsilon[\hat{\mathbf{f}}(\mathbf{x}, \mathbf{v}) - \hat{\mathbf{f}}(\mathbf{x}, \mathbf{u})]. \tag{9.21}$$

If $\mathbf{u}$ is discontinuous at time τ, we use the limit of $\mathbf{u}$ as t tends to τ from below in this expression.

The next step is to follow the two trajectories from these starting points at time τ to time t_1 using the optimal control $\mathbf{u}$ for both of them. Because the components of $\hat{\mathbf{f}}$ are continuously differentiable, the distance between the two trajectories will remain small and $\hat{\boldsymbol{\xi}}(t)$ satisfies the differential equation

$$\dot{\hat{\boldsymbol{\xi}}} = \hat{\mathbf{f}}(\mathbf{y}, \mathbf{u}) - \hat{\mathbf{f}}(\mathbf{x}, \mathbf{u}) = \frac{\partial \hat{\mathbf{f}}(\mathbf{x}, \mathbf{u})}{\partial \hat{\mathbf{x}}} \hat{\boldsymbol{\xi}} = \mathbf{A}\hat{\boldsymbol{\xi}}, \tag{9.22}$$

where $\mathbf{A}$ is the matrix with elements given by

$$A_{ij} = \frac{\partial \hat{f}_i(\mathbf{x}, \mathbf{u})}{\partial \hat{x}_j}. \tag{9.23}$$

At time t_1, the perturbed trajectory reaches the point $\hat{\mathbf{y}}(t_1)$, where

$$\hat{\mathbf{y}}(t_1) = \hat{\mathbf{x}}(t_1) + \hat{\boldsymbol{\xi}}(t_1), \tag{9.24}$$

and $\hat{\boldsymbol{\xi}}(t_1)$ can be found by solving (9.22) with the initial condition given by (9.21).

What we have done is illustrated schematically in Fig. 9.1. The optimal trajectory follows the path SP′PT, using the optimal control $\mathbf{u}$. For times between $\tau - \varepsilon$ and τ we use a different, constant, control $\mathbf{v}$ and move from P′ to Q instead of from P′ to P. The directed element PQ is equal to $\hat{\boldsymbol{\xi}}(\tau)$ and is given by (9.21). For times between τ and t_1 we use the optimal control $\mathbf{u}$ along both paths, ending at time t_1 at the points T and R, and the directed element TR is $\hat{\boldsymbol{\xi}}(t_1)$.

We have considered a single perturbation from the optimal solution. We can find other perturbations by changing the value of the perturbed control $\mathbf{v}$ and the time τ at which we make the perturbation. Each of these will give rise to a final displacement TR, and, since the equation for $\boldsymbol{\xi}$ is linear, we can superimpose these displacements. The set of all possible values of $\hat{\boldsymbol{\xi}}(t_1)$ forms a cone with vertex at $\hat{\mathbf{x}}^1$, which we call the *perturbation cone*. Since the sum of two perturbations is also a possible perturbation the cone is convex. Because the cone is formed by the union of an infinite number of rays, the bounding generators of the cone do not necessarily belong to the perturbation cone. Let $\hat{\mathbf{j}}$ be the unit vector in the direction of the negative x_0 axis. There are a number of possibilities for the position of the perturbation cone relative to $\hat{\mathbf{j}}$, which are illustrated in the two-dimensional case in Fig. 9.2 (in two dimensions the cone becomes a fan). In case 1 the cone and the downward vertical lie on opposite sides of a plane through the vertex of the cone, as in Figs 9.2(a, b); the normal to this separating plane will point downwards, that is, it will have a negative x_0 component. In case 2, $\hat{\mathbf{j}}$ is a generator of the cone and may or may not belong to

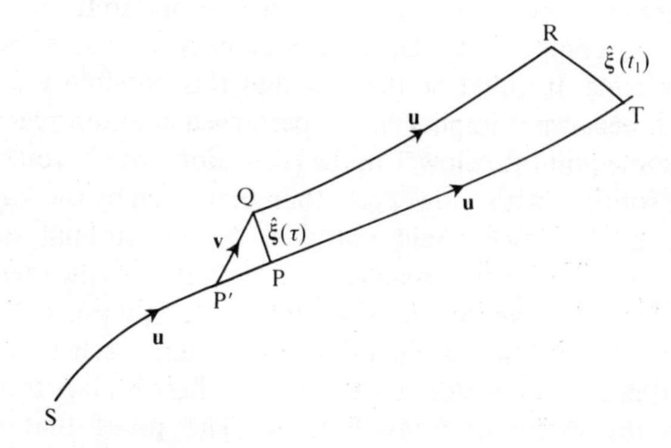

Fig. 9.1. SP′PT is the optimal path and SP′QR is the perturbed path.

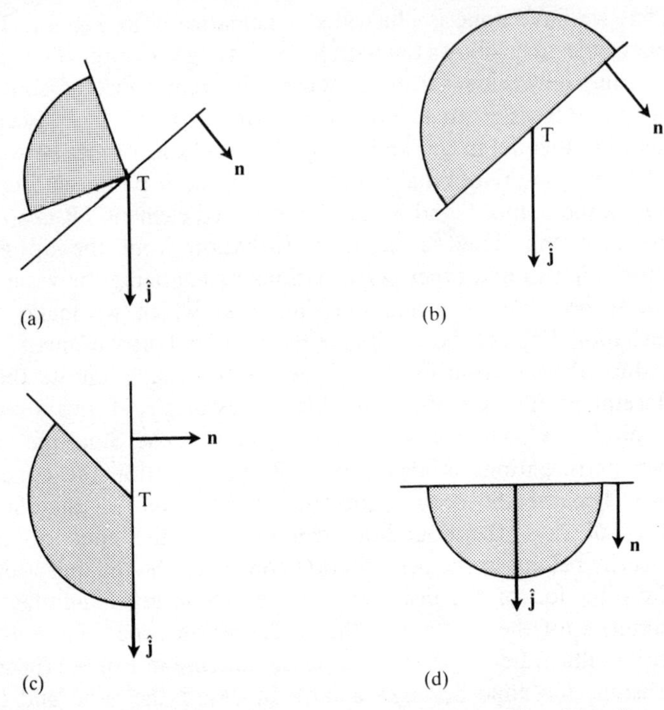

Fig. 9.2. In (a) and (b) the perturbation fan can be separated from $\hat{\mathbf{j}}$ by a line with a downward-pointing normal. In (c), $\hat{\mathbf{j}}$ lies on the boundary of the fan and the normal is perpendicular to $\hat{\mathbf{j}}$. In (d), $\hat{\mathbf{j}}$ lies inside the fan.

the cone itself; see Fig. 9.2(c). The cone now lies on one side of a separating plane and $\hat{\mathbf{j}}$ lies in the plane; the normal to the plane now has a zero x_0 component. The third possibility, case 3, is that $\hat{\mathbf{j}}$ lies within the cone. It might be thought that this possibility could be eliminated, because it implies that a perturbed solution reaches the target at some point R below T in the $\hat{\mathbf{j}}$ direction, which would give a successful solution with a lower cost than that given by the trajectory that ends at T, which could not then be the optimal solution. However, this argument assumes the identity of the perturbed solution and a possible true solution of the state equations. We know that there is no solution of the full problem that reaches a point R below T; it does not immediately follow that there is no perturbation vector in the direction from T to R. The proof that such a perturbation vector does not exist is central to the argument but unfortunately is not easy to prove, even though intuitively it seems quite likely to be true. For the present let us assume that this third possibility for the position of $\hat{\mathbf{j}}$ relative to the perturbation cone cannot occur and postpone further consideration of case 3 to §9.6.

We are left with the first two possibilities, in which there is a separating plane with normal $\hat{\mathbf{n}}$, say, and $\hat{\mathbf{n}} \cdot \hat{\mathbf{j}} > 0$ in case 1, illustrated in Fig. 9.2(a, b), while $\hat{\mathbf{n}} \cdot \hat{\mathbf{j}} = 0$ in case 2; see Fig. 9.2(c). Note that, in Fig. 9.2(a), we could choose a separating plane for which $\hat{\mathbf{n}} \cdot \hat{\mathbf{j}} = 0$, but we can insist that the strict inequality be satisfied. What we have described can be stated formally as follows. There is a hyperplane through the point $\hat{\mathbf{x}}^1$ such that all rays in the perturbation cone lie on one side of this plane or in it, and the vector $\hat{\mathbf{j}}$ lies on the other side of this plane or in it, as illustrated in Figs 9.2(a, b, c). Hence, if the normal to the plane is $\hat{\mathbf{n}}$,

$$\hat{\mathbf{n}}^{\mathrm{T}}\boldsymbol{\xi}(t_1) \leq 0, \tag{9.25}$$

and

$$\hat{\mathbf{n}}^{\mathrm{T}}\hat{\mathbf{j}} > 0 \quad \text{(case 1)}, \quad \hat{\mathbf{n}}^{\mathrm{T}}\hat{\mathbf{j}} = 0 \quad \text{(case 2)}. \tag{9.26}$$

We can now introduce the co-state variables and evaluate the Hamiltonian. For the present we concentrate on case 1, and deal with case 2 in §9.5.

9.4. The co-state variables and the maximum principle: case 1

The co-state equations are part of (9.17), and with H given by (9.16) they can be written in component form as

$$\dot{\hat{z}}_i = -\hat{z}_j \frac{\partial \hat{f}_j(\mathbf{x}, \mathbf{u})}{\partial \hat{x}_i} = -\hat{z}_j A_{ji}, \tag{9.27}$$

where A_{ij} is defined in (9.23). Hence

$$\dot{\hat{\mathbf{z}}} = -\mathbf{A}^{\mathrm{T}}\hat{\mathbf{z}}, \tag{9.28}$$

which is the equation adjoint to the equation (9.22) for $\boldsymbol{\xi}$.

As an initial condition for (9.28), we specify the value of $\hat{\mathbf{z}}(t_1)$, or $\hat{\mathbf{z}}^1$, to be parallel to the normal $\hat{\mathbf{n}}$ to the separating hyperplane defined in (9.26). Since

$$\hat{\mathbf{z}}^1 = \begin{bmatrix} z_0^1 \\ \mathbf{z}^1 \end{bmatrix}, \quad \hat{\mathbf{j}} = \begin{bmatrix} -1 \\ \mathbf{0} \end{bmatrix}, \tag{9.29}$$

the first condition in (9.26) (case 1) shows that $z_0^1 < 0$ and, without loss of generality, we can fix its value so that $z_0^1 = -1$, which is the first part of the theorem.

We now consider the perturbation $\hat{\boldsymbol{\xi}}$ produced by changing the control from $\mathbf{u}$ to $\mathbf{v}$ for $\tau - \varepsilon < t < \tau$. At time τ it is given by (9.21) and its value at time t_1 must satisfy the condition (9.25), that is

$$\hat{\mathbf{z}}^{1\mathrm{T}}\hat{\boldsymbol{\xi}}(t_1) \leq 0. \tag{9.30}$$

Since the equations (9.22) and (9.28) are adjoint, the invariant property recalled in §9.1 from §2.4 shows that

$$\hat{\mathbf{z}}(t)^{\mathrm{T}}\hat{\boldsymbol{\xi}}(t) \leq 0 \quad \text{for } \tau \leq t \leq t_1. \tag{9.31}$$

Applying this condition at time τ to the value of $\hat{\boldsymbol{\xi}}(\tau)$ given by (9.21), we see that

$$\hat{\mathbf{z}}^{\mathrm{T}}\hat{\mathbf{f}}(\mathbf{x}, \mathbf{v}) - \hat{\mathbf{z}}^{\mathrm{T}}\hat{\mathbf{f}}(\mathbf{x}, \mathbf{u}) \leq 0. \tag{9.32}$$

In terms of the Hamiltonian, this condition becomes

$$H(\hat{\mathbf{x}}, \hat{\mathbf{z}}, \mathbf{v}) \leq H(\hat{\mathbf{x}}, \hat{\mathbf{z}}, \mathbf{u}), \tag{9.33}$$

and we have established the maximum principle (9.18).

The perturbed solutions that we have considered have all terminated at time t_1, the optimal terminal time. When this time is fixed by the statement of the optimal control problem, there is nothing more to be said. But in some cases the terminal time is free to take any finite positive value. In the free-time case, we must also allow perturbations in which we change the terminal time from t_1 to $t_1 + \delta$, say, where δ may be positive or negative. We do not perturb the control, so the trajectory follows the optimal path, but terminates before or after it. The state equations at time t_1 then show that the perturbation vectors associated with this change in the terminal time are, to first order in δ, given by

$$\hat{\boldsymbol{\xi}}(t_1) = \delta\hat{\mathbf{f}}[\mathbf{x}(t_1), \mathbf{u}(t_1)] = \delta\hat{\mathbf{f}}(\mathbf{x}^1, \mathbf{u}^1). \tag{9.34}$$

These perturbation vectors must be added to those in the perturbation cone, as illustrated in Fig. 9.3; since both signs of δ are possible, the supporting hyperplane must include the vector $\hat{\mathbf{f}}(\mathbf{x}^1, \mathbf{u}^1)$. Since $\hat{\mathbf{z}}^1$ is normal to this hyperplane, it follows that

$$H(\hat{\mathbf{x}}^1, \hat{\mathbf{z}}^1, \mathbf{u}^1) = \hat{\mathbf{z}}^{1\mathrm{T}}\hat{\mathbf{f}}(\mathbf{x}^1, \mathbf{u}^1) = 0, \tag{9.35}$$

and we have shown that the Hamiltonian is zero at the terminal time.

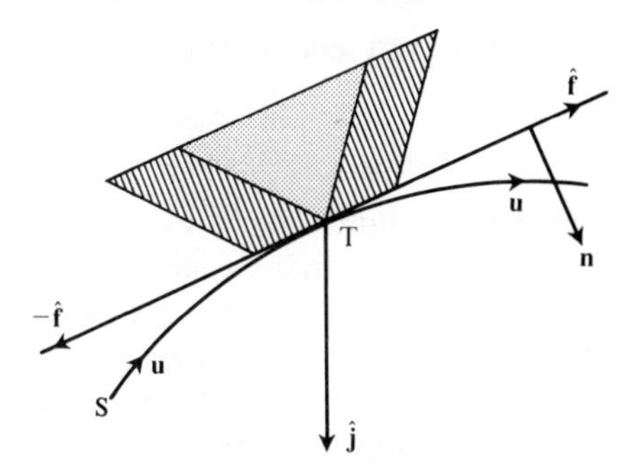

Fig. 9.3. The perturbation cone with the terminal-time perturbations added.

The final result that we have to establish is that the Hamiltonian is constant along the optimal trajectory. The result we have just proved then shows that this constant is zero in the free-time case. When there is no restriction on the magnitude of the controls and the optimal control **u** is regular, the maximum principle shows that, along the optimal trajectory,

$$\frac{\partial H}{\partial \mathbf{u}} = \mathbf{0}. \tag{9.36}$$

It follows that

$$\frac{\mathrm{d}H}{\mathrm{d}t} = \frac{\partial H}{\partial \hat{\mathbf{x}}} \cdot \dot{\hat{\mathbf{x}}} + \frac{\partial H}{\partial \hat{\mathbf{z}}} \cdot \dot{\hat{\mathbf{z}}} = 0, \tag{9.37}$$

where we have used the state and co-state equations (9.17), and H is constant along the optimal trajectory. Since, however, we know that the optimal control is often discontinuous, we need a more refined argument to establish this result in general.

Let t and t' be two times between 0 and t_1, and let us define $p(t, t')$ by

$$p(t, t') = H[\hat{\mathbf{x}}(t), \hat{\mathbf{z}}(t), \mathbf{u}(t')]. \tag{9.38}$$

Then $p(t, t)$ is the value of H at time t along the optimal trajectory, and we wish to prove that $p(t, t) = 0$. From the maximum principle we know that

$$p(t, t) \geq p(t, t'), \quad p(t', t') \geq p(t', t). \tag{9.39}$$

Also, since H is a differentiable function of $\hat{\mathbf{x}}$ and $\hat{\mathbf{z}}$,

$$|p(t', t) - p(t, t)| < \delta, \quad |p(t, t') - p(t', t')| < \delta', \tag{9.40}$$

where δ and δ' are positive and tend to zero as $t' - t$ tends to zero. It follows that

$$-\delta' \leq p(t, t) - p(t', t') \leq \delta. \tag{9.41}$$

Hence $p(t, t)$ is a continuous function of t, even though $\mathbf{u}$ may not be continuous. We know, however, that $\mathbf{u}$ is piecewise continuous, so if we can show that $p(t, t)$ is constant between two successive discontinuities of $\mathbf{u}$, $p(t, t)$ is equal to the same constant for all t, since we have already proved it to be continuous. But $p(t_1, t_1) = 0$ from (9.35) and so the Hamiltonian is zero along the optimal trajectory.

We have to prove that $(\mathrm{d}/\mathrm{d}t)p(t', t') = 0$ when $\mathbf{u}$ is continuous at time t'. Because $\mathbf{u}$ is piecewise continuous, there is an interval including t' in which $\mathbf{u}$ is continuous. Suppose that t lies in this interval. Then, from (9.39),

$$p(t, t) - p(t', t') \leq p(t, t) - p(t', t), \tag{9.42}$$

and

$$p(t, t) - p(t', t') \geq p(t, t') - p(t', t'). \tag{9.43}$$

It follows that, when $t > t'$,

$$\frac{p(t, t) - p(t', t')}{t - t'} \leq \frac{H(\hat{\mathbf{x}}, \hat{\mathbf{z}}, \mathbf{u}) - H(\hat{\mathbf{x}}, \hat{\mathbf{z}}', \mathbf{u})}{t - t'} + \frac{H(\hat{\mathbf{x}}, \hat{\mathbf{z}}', \mathbf{u}) - H(\hat{\mathbf{x}}', \hat{\mathbf{z}}', \mathbf{u})}{t - t'}, \tag{9.44}$$

and

$$\frac{p(t, t) - p(t', t')}{t - t'} \geq \frac{H(\hat{\mathbf{x}}, \hat{\mathbf{z}}, \mathbf{u}') - H(\hat{\mathbf{x}}, \hat{\mathbf{z}}', \mathbf{u}')}{t - t'} + \frac{H(\hat{\mathbf{x}}, \hat{\mathbf{z}}', \mathbf{u}') - H(\hat{\mathbf{x}}', \hat{\mathbf{z}}', \mathbf{u}')}{t - t'}, \tag{9.45}$$

where the primed variables are evaluated at t' and the unprimed ones at t. As t tends to t', the right-hand sides of the two inequalities both tend to zero, since we know from (9.15) and (9.17) that

$$\frac{\partial H}{\partial \hat{\mathbf{z}}} \cdot \dot{\hat{\mathbf{z}}} + \frac{\partial H}{\partial \hat{\mathbf{x}}} \cdot \dot{\hat{\mathbf{x}}} = 0. \tag{9.46}$$

Similar inequalities to (9.44) and (9.45) hold when $t < t'$, so we have established that $\dot{p}$ is zero at t' and hence that p is constant over any interval within which $\mathbf{u}$ is continuous. The proof that H is zero along the optimal trajectory is now complete.

As an illustration of the above result, consider the positioning problem of §6.3(3). The Hamiltonian for that problem is given by

$$H = -1 + z_1 x_2 + z_2 u_1, \tag{9.47}$$

and u_1 is equal to -1 for $0 \leq t < t_2$ and $u_1 = 1$ for $t_2 < t \leq t_1$. Thus the optimal control has a discontinuity at $t = t_2$. In both these time intervals, z_1 and u_1 are constant, $\dot{z}_2 = -z_1$, and $\dot{x}_2 = u_1$, so that H is indeed constant within each interval. The continuity of H at the switching time t_2 follows because z_1 and x_2 are continuous, and the maximum principle shows that z_2 is zero at time t_2, when the optimal control switches from one extreme value to the other. Thus $z_2 u_1$ is also continuous and so H is continuous for $0 \leq t \leq t_1$.

We have now completed the proof of the PMP in case 1. Before we deal with cases 2 and 3, it is convenient to consider the extensions to the PMP described in Chapter 7. These dealt with more general target sets, terminal costs, and non-autonomous problems. The second and third of these extensions were derived from the basic form of the PMP in §7.2 and §7.3. The extension to cover general targets was stated in §7.1 but not proved there. The condition to be applied when the target point is not fixed is that the (unextended) co-state vector must be normal to the target set at the terminal time, that is, $\mathbf{z}^1$ is normal to $\mathscr{T}$ at time t_1. Consider the projection of the perturbation cone in the x_0 direction. The separating plane must have the perturbation cone on one side and the target set on the other, else we could reach the target set at lower cost. This proves that $\mathbf{z}^1$ must be normal to $\mathscr{T}$ at the point where the optimal trajectory reaches the target.

The transversality condition can be extended to cover problems in which the initial state is not completely specified. In the time-reversed system the roles of the initial and final states are interchanged, so that the same condition holds. The initial co-state vector must be normal to the set of initial state vectors; that is, if the initial state must lie in a set $\mathscr{I}$, $\mathbf{z}^0$ must be normal to $\mathscr{I}$ at the point where the optimal trajectory leaves the initial state.

9.5. Case 2

We consider now the possibility that the boundary of the perturbation cone includes the vector $\hat{\mathbf{j}}$, as in Fig. 9.2(c). From (9.26) this means that we must have $z_0^1 = 0$ and can no longer normalize the co-state variable so that $z_0^1 = -1$. The remainder of the argument leading to the PMP is unaltered, so it appears that the statement of the theorem in §6.2 and in §9.2 is erroneous. However, in all the examples we have solved, and in all cases known to me, the change in the condition on z_0^1 leads either to the optimal solution found when z_0^1 is equal to -1 or to no solution. The first of these alternatives corresponds to the situation illustrated in Fig. 9.2(a) when we could choose the normal $\hat{\mathbf{n}}$ so that $\hat{\mathbf{n}} . \hat{\mathbf{j}} = 0$, but it is not necessary to do so. Hence, although there does not seem to be any reason why the possibility that $z_0^1 = 0$ should be excluded, in practice it does not seem to occur. One reason for the failure of this possibility to yield a solution is that, since z_0 is known to be constant, the term in the Hamiltonian depending on the cost function is absent. Any optimal solution produced by applying this form of the condition would therefore have to be optimal whatever cost function were being employed. But the possibility that in some exceptional circumstances we might have the condition $z_0^1 = 0$ cannot be ruled out.

To illustrate the procedure when this condition is employed, let us consider again the positioning problem of §6.3(3), for which we found an optimal solution using the condition $z_0^1 = -1$. With $z_0^1 = 0$, the solution proceeds as before, and we again find that the control must be bang–bang. But now, when we apply the condition that H is zero along the optimal trajectory we find that $z_2^1 = 0$, so that, from (6.52),

$$z_2 = A(t_1 - t) \tag{9.48}$$

and z_2 cannot change sign. Hence the only solutions we can find are those for which no switching is required, that is, when the initial state lies on the switching curve SOS of Fig. 6.1. But these initial states can be dealt with by the former condition, so that no additional solutions are provided by extending the condition on z_0^1 to allow it to take the value 0.

9.6. Case 3

We return now to the problem of excluding the case when the perturbation cone contains the downward-pointing vector $\hat{\mathbf{j}}$ in its interior, as illustrated in Fig. 9.2(d). The difficulty arises because of the necessity to distinguish between the solutions of the state equations (possibly nonlinear) and the solutions arising from a perturbation of the optimal solution of the kind we have been considering in §9.3. Consider the possible configuration for a two-dimensional problem sketched in Fig. 9.4.

The point T is the optimal target point and the shaded regions cover those points close to T that can be reached by an acceptable choice of controls. Since the control leading to T is optimal the reachable shaded regions cannot include the line below T. The infinitesimal perturbation vectors, extended to a length equal to some small positive ε, have one end at T and the other at points on the semicircle as shown. Since the sum of two perturbations is also a perturbation there are no gaps in the semicircle. Now suppose that P is any point of the semicircle; the perturbation to the optimal solution that leads to the perturbation vector TP produces a solution to the complete state equations (not their linearized forms) that terminates in the point R, say, and the distance between P and R is $o(\varepsilon)$. Suppose there is one perturbation that leads to the pair of points P_1 and R_1 and another that leads to P_2 and R_2. Because the distance of R_1 from P_1 is $o(\varepsilon)$ we can ensure that R_1 is in the left-hand section of the reachable states and similarly that R_2 is in the right-hand section. Now consider a sequence of perturbations that moves P from P_1 to P_2 continuously. Then R will move continuously from R_1 to R_2 and must therefore cross the line below T, which is a contradiction. This argument can be made more precise and can be extended to the general n-dimensional case, but it requires the application of a variant of the Brouwer fixed-point theorem. In higher dimensions, it is possible to move from R_1 to R_2 without crossing the line below T, so a more sophisticated argument than the one outlined for the two-dimensional case must be used.

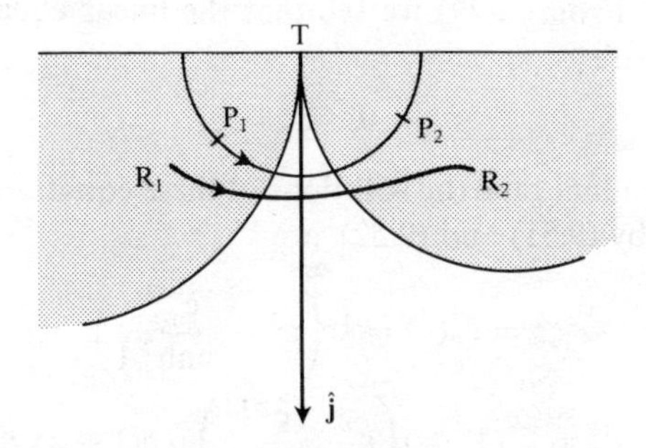

Fig. 9.4. The perturbation cone includes the downward vertical.

This completes our discussion of a way by which the Pontryagin maximum principle can be established. We conclude this chapter by carrying out the steps involved in the proof of the PMP for a specific example, for which we can calculate the perturbation cone.

9.7. An illustrative example of the proof of the PMP

Consider the system with cost and state equations given by

$$\dot{x}_0 = f_0(x_1, u_1) = \tfrac{1}{2}u_1^2, \quad \dot{x}_1 = f_1(x_1, u_1) = x_1 + u_1, \tag{9.49}$$

and with the initial and final states $x_1(0) = 0$, $x_1(1) = 1$. The target and the terminal time are both fixed, and the optimal solution has the form

$$x_0 = \frac{1 - \mathrm{e}^{-2t}}{4\sinh^2 1}, \quad x_1 = \frac{\sinh t}{\sinh 1}, \quad z_1 = u_1 = \frac{\mathrm{e}^{-t}}{\sinh 1},$$
$$J = x_0(1) = \frac{1}{\mathrm{e}^2 - 1}. \tag{9.50}$$

Following the steps of the argument presented in §9.3, we find the perturbation cone. We pick a time τ between 0 and 1 and for the small time interval from $\tau - \varepsilon$ to τ we change the optimal control u_1 to another control v. The new state of the system at time τ is $x_0(\tau) + \xi_0(\tau)$ and $x_1(\tau) + \xi_1(\tau)$ and

$$\xi_0(\tau) = \varepsilon[f_0(x_1, v) - f_0(x_1, u_1)] = \tfrac{1}{2}\varepsilon\left(v^2 - \frac{\mathrm{e}^{-2\tau}}{\sinh^2 1}\right), \tag{9.51}$$

$$\xi_1(\tau) = \varepsilon[f_1(x_1, v) - f_1(x_1, u_1)] = \varepsilon\left(v - \frac{\mathrm{e}^{-\tau}}{\sinh 1}\right). \tag{9.52}$$

Starting from this perturbed state, we now apply the optimal control for $\tau < t < 1$. From (9.49) we see that the linear equations for the perturbation are

$$\dot{\xi}_0 = 0, \quad \dot{\xi}_1 = \xi_1, \tag{9.53}$$

and at the terminal time the solutions of these equations, with initial values given by (9.51) and (9.52), are

$$\xi_0^1 = \xi_0(1) = \tfrac{1}{2}\varepsilon\left(v^2 - \frac{\mathrm{e}^{-2\tau}}{\sinh^2 1}\right),$$
$$\xi_1^1 = \xi_1(1) = \varepsilon\left(v - \frac{\mathrm{e}^{-\tau}}{\sinh 1}\right)\exp(1 - \tau). \tag{9.54}$$

Since it follows from these results that

$$\xi_0^1 - \xi_1^1 \frac{e^{-1}}{\sinh 1} = \tfrac{1}{2}\varepsilon\left(v - \frac{e^{-\tau}}{\sinh 1}\right)^2, \tag{9.55}$$

all perturbation vectors found by varying v and τ lie above the line through the target point with slope $e^{-1}/\sinh 1$. Adding together different perturbation vectors yields another perturbation vector on the same side of this line, so the perturbation cone is the semicircle shown in Fig. 9.5.

Now that we have identified the perturbation cone we can proceed, as in §9.4, to find the co-state vector. The diameter of the perturbation semicircle lies on the separating line, and we take the co-state vector at the terminal time along the normal to this line, so that

$$z_0^1 = -1, \quad z_1^1 = \frac{e^{-1}}{\sinh 1}. \tag{9.56}$$

The adjoint equations are

$$\dot{z}_0 = 0, \quad \dot{z}_1 = -z_1, \tag{9.57}$$

and the solution that satisfies the conditions (9.56) at time $t = 1$ has values at time $t = \tau$ given by

$$z_0(\tau) = -1, \quad z_1(\tau) = \frac{e^{-\tau}}{\sinh 1}. \tag{9.58}$$

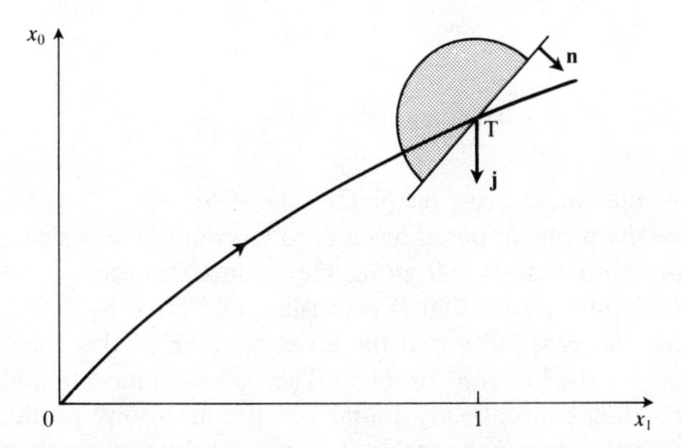

Fig. 9.5. The perturbation cone is the shaded semicircle. The diameter lies on the line separating the perturbation cone from the downward line through the target point.

At the terminal time, we know that the perturbation vector and the co-state vector are on opposite sides of the separating line, so that

$$z_0^1\xi_0^1 + z_1^1\xi_1^1 \le 0, \tag{9.59}$$

and the same inequality holds for all times, by the property of adjoint systems. Hence, from (9.51) and (9.52),

$$z_0(\tau)[f_0(x_1, v) - f_0(x_1, u_1)] + z_1(\tau)[f_1(x_1, v) - f_1(x_1, u_1)] \le 0, \tag{9.60}$$

which can be rewritten in the form

$$z_0 f_0(x_1, v) + z_1 f_1(x_1, v) \le z_0 f_0(x_1, u_1) + z_1 f_1(x_1, u_1) \tag{9.61}$$

or

$$H(x_1, z_0, z_1, v) \le H(x_1, z_0, z_1, u_1), \tag{9.62}$$

which is the PMP. From the known optimal trajectory and co-state variables we can calculate these Hamiltonians. Since

$$H(x_1, z_0, z_1, u_1) = \tfrac{1}{2}u_1^2 z_0 + (x_1 + u_1)z_1, \tag{9.63}$$

we find that, at time τ,

$$H(x_1, z_0, z_1, v) = -\frac{1}{2}\left(v - \frac{e^{-\tau}}{\sinh 1}\right)^2 + \frac{1}{2\sinh^2 1}, \tag{9.64}$$

and

$$H(x_1, z_0, z_1, u_1) = \frac{1}{2\sinh^2 1}, \tag{9.65}$$

and the maximum principle (9.62) is satisfied.

Since the problem posed has a fixed terminal time, we do not use the condition that $H = 0$ along the optimal trajectory; it follows from the above results that H is constant and given by (9.65). Let us consider the possibility that the given solution is also the optimal solution for the free-time problem. Then we shall have to add to the perturbation cone already found the perturbations produced by changing the terminal time to $1 \pm \delta$, for arbitrary small positive values of δ. From the known values of x_0 and x_1 given by (9.50), the perturbations η_0 and η_1 produced by this change in the terminal time are given to first order in δ by

$$\eta_0 = \frac{1 - \exp(-2 \pm 2\delta)}{4 \sinh^2 1} - \frac{1 - \exp(-2)}{4 \sinh^2 1} = \pm\delta \frac{\exp(-2)}{2 \sinh^2 1}, \quad (9.66)$$

$$\eta_1 = \frac{\sinh(1 \pm \delta)}{\sinh 1} - 1 = \pm\delta \coth 1. \quad (9.67)$$

These perturbations must be added to those forming the perturbation cone of Fig. 9.5 and together they produce perturbations lying in the shaded region in Fig. 9.6. We see that there are possible perturbations in all directions from the target point and the total perturbation cone is therefore the whole disc centred at the target including the downward-pointing line. Hence there are perturbations that lead to a cost lower than the optimal cost when the terminal time is $t_1 = 1$.

There are solutions to the full problem, and not only of its linearized version, that lower the cost. If the terminal time is $1 + \delta$ and we use the control

$$u_1' = \frac{e^{-t}}{\sinh(1 + \delta)}, \quad (9.68)$$

we can arrive at the target $x_1(1) = 1$ at the terminal time, the new trajectory and cost being given by

$$x_1' = \frac{\sinh t}{\sinh(1 + \delta)}, \quad J' = \frac{1}{\exp(2 + 2\delta) - 1}, \quad (9.69)$$

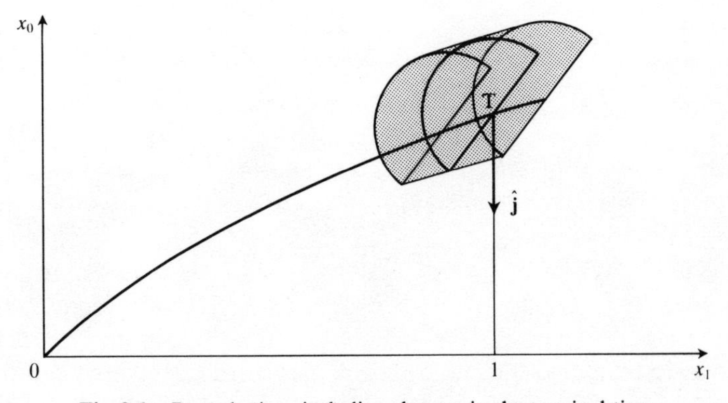

Fig. 9.6. Perturbations including changes in the terminal time.

and $J' < J$, the cost given in (9.50), when $\delta > 0$. In fact, the solution given by (9.68) and (9.69) holds for an arbitrary terminal time $t_1 = 1 + \delta$ with δ not necessarily small, so the cost can be made arbitrarily small by taking the terminal time large enough. The free-time problem has no optimal solution.

We have now established the truth of the PMP, which we used in Chapters 6, 7, and 8 to solve a number of problems. In the next chapter we apply the PMP to some more complicated examples.

Exercises 9

(*Solutions on p. 236*)

9.1 Use the PMP to solve one of the time-optimal problems described in Chapters 4 and 5, for example the harmonic oscillator (see §5.3).

9.2 In the problem discussed in §6.4, the optimal control was found to be discontinuous at two intermediate times. Evaluate the Hamiltonian and show that, in spite of the discontinuities in the control, H is continuous along the optimal trajectory.

9.3 Find the optimal solution of the following problem and illustrate the proof of the PMP by finding the perturbation cone and the co-state vector:

state equation $\dot{x}_1 = x_1 + u_1$; bounded control $|u_1| \leq 1$;
initial state $x_1(0) = 0$; final state $x_1(t_1) = 1$; cost function

$$J = \int_0^{t_1} 1 \, dt.$$

10 Further applications and extensions

> Example is always more efficacious than precept.
>
> Samuel Johnson

This chapter contains some further applications of the PMP of a more difficult nature. Now that some experience in using the PMP for simple problems has been gained, the solutions presented here are not given in as much detail as before. We also consider more general control sets and the way in which restrictions on the state variables can be included in the theory.

10.1. The Moon-landing problem

The name given to this problem is explained by the planning of the succesful landing of a module on the surface of the Moon in the *Surveyor* missions of 1966. The descent of the module had to be controlled so that the landing was 'soft', that is, both the height of the module above the surface and its downward velocity had to be brought to zero simultaneously. Because of the limited amount of fuel that could be carried on the module, it was important to achieve the soft landing with as small an expenditure of fuel as possible. The time taken to reach the surface was of little significance. A simplified version of this problem is described and solved in this section.

A module of mass M is initially at a height h above the surface of the Moon and is moving with a downward velocity V. It falls under a uniform gravitational acceleration g and there is no atmosphere to resist its motion. Its velocity can be controlled by engines that can exert a maximum thrust F in either the upward or downward direction. The target is the surface of the Moon, which must be reached with zero velocity, and the cost function is the total fuel consumed, which is proportional to the time integral of the magnitude of the thrust exerted by the engines. The system can be defined at any time τ by the height $y(\tau)$ of the module above the surface and its upward velocity $dy/d\tau$. The control variable is the thrust exerted by the engines, say $f(\tau)$, and $|f(\tau)| \leq F$. The state equations are then

$$M\frac{\mathrm{d}^2 y}{\mathrm{d}\tau^2} = -Mg + f, \tag{10.1}$$

and the initial and final states are given by

$$y = h, \frac{\mathrm{d}y}{\mathrm{d}\tau} = -V, \quad \text{at } \tau = 0; \quad y = 0, \frac{\mathrm{d}y}{\mathrm{d}\tau} = 0, \quad \text{at } \tau = \tau_1, \tag{10.2}$$

where τ_1 is the terminal time. The cost is given by

$$K = \int_0^{\tau_1} |f(\tau)|\mathrm{d}\tau. \tag{10.3}$$

We define new variables to reduce the problem to the standard form. Let

$$\tau = \left[\frac{M}{F}\right]^{1/2} t, \quad y = x_1, \quad \frac{\mathrm{d}y}{\mathrm{d}\tau} = \left[\frac{F}{M}\right]^{1/2} x_2, \quad f = Fu_1,$$
$$K = (MF)^{1/2} J. \tag{10.4}$$

The state equations and conditions then have the form

$$\dot{x}_0 = |u_1|, \quad \dot{x}_1 = x_2, \quad \dot{x}_2 = -c + u_1, \tag{10.5}$$

$$x_0(0) = 0, \quad x_1(0) = h, \quad x_2(0) = -v, \tag{10.6}$$

$$x_0(t_1) = J, \quad x_1(t_1) = 0, \quad x_2(t_1) = 0. \tag{10.7}$$

In these equations

$$t_1 = \left[\frac{F}{M}\right]^{1/2} \tau_1, \quad v = \left[\frac{M}{F}\right]^{1/2} V, \quad c = \frac{Mg}{F} < 1, \quad |u_1| \leq 1. \tag{10.8}$$

The reason for the upper bound on the value of c is that if $c \geq 1$ it is impossible to slow down the descent of the module and make a soft landing. Additionally, we must drop on to the surface from above, which means that we approach the target from within the quadrant $x_1 > 0$, $x_2 < 0$. Indeed, we must have $x_1 > 0$ throughout the motion.

The Hamiltonian for this problem is given by

$$H = -|u_1| + z_1 x_2 + z_2(u_1 - c), \tag{10.9}$$

and the co-state equations are

$$\dot{z}_1 = 0, \quad \dot{z}_2 = -z_1, \tag{10.10}$$

which have solutions of the form

$$z_1 = A, \quad z_2 = B - At. \tag{10.11}$$

As in the fuel-cost problem of §6.4, the supremum of H is achieved by taking $u_1 = 0$ when $|z_2| < 1$, $u_1 = 1$ when $z_2 > 1$, and $u_1 = -1$ when $z_2 < -1$. On the approach to the target we must have $x_2 < 0$ and, since we end with $x_2 = 0$, we must have $\dot{x}_2 > 0$. Hence, in the final stage we use the control $u_1 = 1$ and the corresponding part of the trajectory must be given by

$$x_1 = \tfrac{1}{2}(1-c)(t-t_1)^2, \quad x_2 = (1-c)(t-t_1), \quad x_2^2 = 2(1-c)x_1. \tag{10.12}$$

The initial state must lie on or to the right of this curve; otherwise the trajectory is bound to reach the surface $x_1 = 0$ with $x_2 < 0$, even when the engines are producing their maximum upward thrust. Thus for a successful soft landing, the initial state must satisfy the condition

$$v^2 \leq 2(1-c)h; \tag{10.13}$$

see Fig. 10.1.

Except when $v^2 = 2(1-c)h$, this final part of the trajectory must be preceded by a part for which the control $u_1 = 0$, since the continuity of z_2 ensures that we must not switch directly from $+1$ to -1.

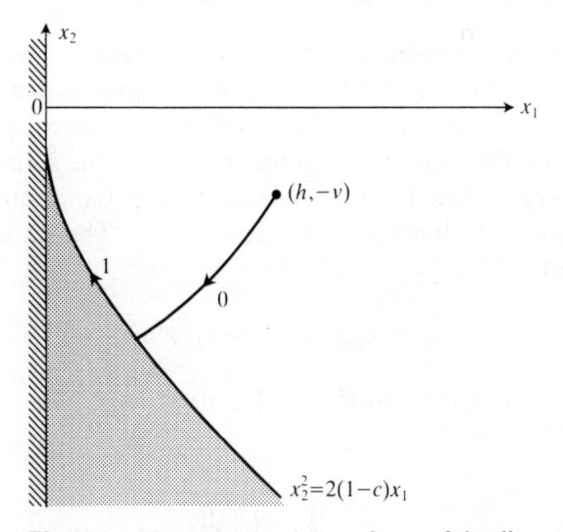

Fig. 10.1. The optimal trajectory for a soft landing.

A second switch from 0 to -1 will not be needed if the initial value of $z_2 \geq -1$, that is, if $B \geq -1$. The trajectory through the initial state, using the control $u_1 = 0$, is given by

$$x_1 = h - vt - \tfrac{1}{2}ct^2, \quad x_2 = -v - ct. \tag{10.14}$$

If the switch occurs at $t = t_2$, where $0 < t_2 < t_1$, we can equate the values of x_1 and x_2 given by (10.12) and (10.14), and we can also use the condition that $H = 0$ at $t = t_1$, say. These equations provide enough information to solve the problem and we find that

$$J = \left[\frac{2hc + v^2}{1 - c}\right]^{1/2}, \quad B = \frac{v}{v + ct_2} > 0. \tag{10.15}$$

The values of t_2 and t_1 can also be found. Since $B > -1$, there is no need to begin the descent with a period during which $u_1 = -1$. Such an initial firing of the engines would only be appropriate if, initially, the module were moving upwards. For the case under consideration, the optimal strategy is to allow the module to fall downwards under gravity until, at a critical height above the surface, the engines are fired to produce their maximum upward thrust and the module lands on the surface with zero speed at impact. This sequence of controls minimizes the expenditure of fuel.

10.2. The harmonic oscillator with a fuel cost

The time-optimal problem for the harmonic oscillator was discussed in §5.3 using the methods of Part A. For the same system with a cost that measures the amount of fuel used in the control process, the PMP must be the tool by which we can obtain the solution. In the time-optimal problem, it was found that a bang–bang control should be employed, with switchings at intervals of π. The state equations have the form

$$\dot{x}_1 = x_2, \quad \dot{x}_2 = -x_1 + u_1, \tag{10.16}$$

with bounded controls, so that $-1 \leq u_1 \leq +1$. The cost is measured by

$$J = \int_0^{t_1} |u_1| \, dt. \tag{10.17}$$

The terminal time t_1 is free to take any positive value, and the initial and final states are given by

$$x_1(0) = X_1, \quad x_2(0) = X_2, \quad x_1(t_1) = 0, \quad x_2(t_1) = 0. \quad (10.18)$$

Let J_{T} be the optimal cost for the time-optimal problem, with terminal time $t_1 = T$. The optimal controls are bang–bang and it is clear that they provide successful controls for the fuel-cost problem as well, with a cost equal to J_{T}. Hence the optimal solution J_{F} of the fuel-cost problem is less than or equal to J_{T}. Also, if the terminal time in the optimal solution of the fuel-cost problem is less than T, we should have found an admissible control sequence that enables us to reach the target in less time than the optimal one, which is impossible. Hence the optimal cost for the fuel problem is less than that for the time problem, but the terminal time is increased. From the definition (10.17) of the cost of the fuel, we see that this can only be achieved if the control is not bang–bang.

To solve the problem, we form the Hamiltonian H, given by

$$H = -|u_1| + z_1 x_2 + z_2(-x_1 + u_1), \quad (10.19)$$

and the supremum of H will be achieved if we choose

$$u_1 = 1 \text{ when } z_2 > 1, \quad u_1 = 0 \text{ when } |z_2| < 1,$$
$$u_1 = -1 \text{ when } z_2 < -1. \quad (10.20)$$

The co-state equations have the form

$$\dot{z}_1 = z_2, \quad \dot{z}_2 = -z_1, \quad (10.21)$$

and the point (z_1, z_2) must lie on a circle centre $(0, 0)$ and radius R, say. If $R \leq 1$, we must have $u_1 = 0$ for all t, and it is impossible to reach the target. It follows that the control must follow the sequence $\{1, 0, -1, 0, 1, \dots\}$. Since the path of $\mathbf{z}$ is traced out at unit angular rate, and since the switchings occur whenever z_2 passes through the values ± 1, it follows that the controls ± 1 are applied for equal times α, say, and the zero control is applied for equal times $\pi - \alpha$, where $0 < \alpha < \pi$. The corresponding trajectories are composed of arcs of circles in the (x_1, x_2) plane; the centres of these circles are at $(1, 0)$ when $u_1 = 1$, at $(0, 0)$ when $u_1 = 0$, and at $(-1, 0)$ when $u_1 = -1$. The solution for a general initial state can be found by piecing together a succession of circular arcs. As a specific example, consider the initial state $X_1 = 0, X_2 = 2$. The control sequence

$\{-1, 1\}$ steers this point to the origin via the point $(1, -1)$ in a total time π, with cost also equal to π. This is, in fact, the strategy for the time-optimal problem. The sequence $\{0, 1\}$ of controls is also successful, with the switch occurring when the state is $(2, 0)$. The cost is again equal to π and the terminal time is $\frac{3}{2}\pi$, but this cannot be the optimal solution since the total time during which the pair of controls is used is greater than π. If we now try the sequence of controls $\{-1, 0, 1\}$ and piece together the corresponding arcs, we find that the first switch must be made at time $2\alpha - \frac{1}{2}\pi$, the second at 2α, and the target is reached at time $2\alpha + \frac{1}{2}\pi$, where $\alpha = \tan^{-1} 2$. The optimal cost is equal to the total time during which a non-zero control is being used, and this is equal to 2α. Note that this is less than the time-optimal cost, which is equal to π, but the terminal time is greater than π, as anticipated. The trajectory of this optimal solution is sketched in Fig. 10.2.

It must be remembered, however, that the PMP only provides necessary conditions for the optimal solution. It may be that there are many solutions that are satisfactory as far as the PMP is concerned. If so, we must find them all and then pick the one that

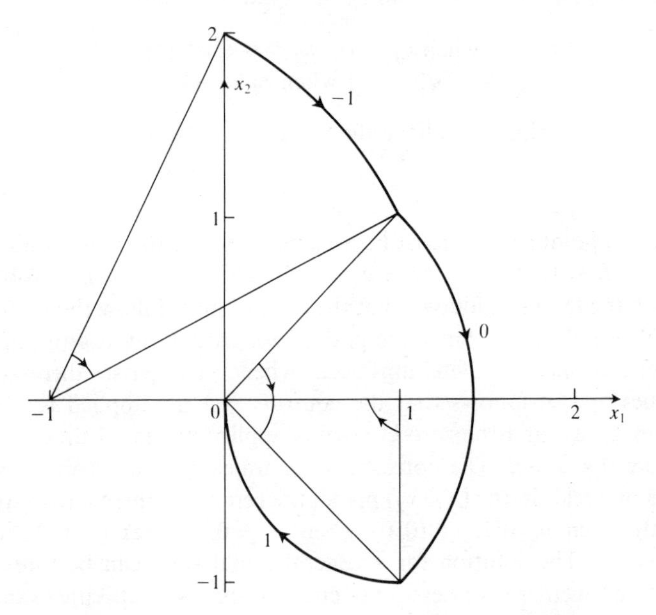

Fig. 10.2. An optimal solution for the harmonic oscillator with fuel cost.

gives the lowest cost. In the present example there are, in fact, many acceptable solutions. Suppose the optimal trajectory from the given initial point first crosses the x_1 axis at the point $x_1 = X$. Now we apply the following cycle of controls: $u_1 = 0$ for a time α, $u_1 = -1$ for a time $\pi - 2\alpha$, and $u_1 = 0$ for a further time α. On solving the state equations we find that after the total time π we reach the point $x_1 = 2\cos\alpha - X$, $x_2 = 0$, and the cost incurred is equal to $\pi - 2\alpha$. If we repeat this cycle for a further time π, but with the sign of u_1 reversed, we reach the point $x_1 = X - 4\cos\alpha$, $x_2 = 0$ at a total cost equal to $2(\pi - 2\alpha)$. It follows that we can reach the target after n such cycles, provided we choose α so that $\cos\alpha = X/2n$, and that the total cost along the trajectory from $(X, 0)$ to $(0, 0)$ is given by

$$J_n = n\left[\pi - 2\cos^{-1}\left(\frac{X}{2n}\right)\right] = 2n\sin^{-1}\left(\frac{X}{2n}\right). \tag{10.22}$$

Although the total time increases with n, the cost decreases and, for large values of n,

$$J_n = X + \frac{X^3}{24n^2} + \cdots. \tag{10.23}$$

We conclude, therefore, that there is no optimal solution, since the set of values of J_n is bounded below by X, but is never equal to X.

10.3. Positioning with smoothly varying controls

In the standard positioning problem, we allow the controls to switch instantaneously from one extreme value to the other. This may not always be possible or indeed desirable, because of the stresses produced by a rapid change of acceleration. As a variant of the problem, we can consider a system under the control of forces whose values can be increased or decreased at a maximum permitted rate. Such a system has already been considered in the exercises following Chapter 7, but here we justify the equations stated there and solve a problem with more realistic terminal conditions.

We can derive the equations for a system with these characteristics as follows. As depicted in Fig. 10.3, we have a wheeled trolley which is able to move freely in a straight horizontal line and its motion is controlled by strings in both directions attached to two cups. The masses on either side can be increased by pouring sand into the cups at a given maximum rate.

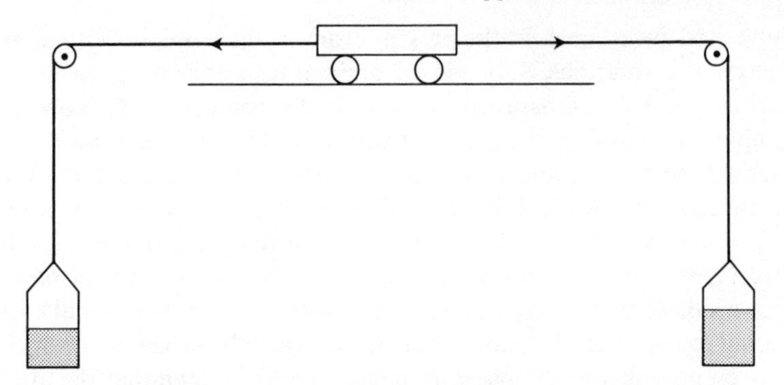

Fig. 10.3. Trolley controlled by smoothly varying forces.

If the mass M of the trolley is much larger than the other masses, we can express the equations of motion for the trolley in the form

$$\dot{x}_1 = x_2, \quad M\dot{x}_2 = (m_1 - m_2)g, \tag{10.24}$$

where x_1 and x_2 are the position and velocity of the trolley, m_1 and m_2 the masses on either side, and g the gravitational acceleration. Because the mass of the trolley is so large, we can neglect the rate of change of the momentum of the cups and sand. The masses m_1 and m_2 can be changed by the addition of sand, so that

$$\dot{m}_1 = v_1, \quad \dot{m}_2 = v_2, \tag{10.25}$$

where $0 \leq v_1 \leq V$ and $0 \leq v_2 \leq V$. If we define new variables x_3 and u_1 by

$$x_3 = \frac{(m_1 - m_2)g}{M}, \quad u_1 = \frac{(v_1 - v_2)g}{M}, \tag{10.26}$$

the system is specified by the state equations

$$\dot{x}_1 = x_2, \quad \dot{x}_2 = x_3, \quad \dot{x}_3 = u_1, \tag{10.27}$$

and $-h \leq u_1 \leq h$, where $h = Vg/M$.

For a specific problem let us suppose that initially the trolley is at rest at a positive distance X from the origin and that $m_1 = m_2$. The target is the origin, so that the initial and final conditions are given by

$$x_1(0) = X, \quad x_2(0) = 0, \quad x_3(0) = 0, \quad x_1(t_1) = 0, \quad x_2(t_1) = 0, \tag{10.28}$$

where t_1 is the terminal time. In this problem the goal is to reach the desired position with zero speed there. If additionally we require that $x_3(t_1) = 0$, we can set $u_1 = 0$ for $t > t_1$ and the trolley can remain at rest at the origin. To complete the specification of the problem, we need a cost function. We suppose that the target is to be reached quickly, but that also a small amount of sand is to be used in the controls. Thus we may define the cost function by

$$J = \int_0^{t_1} (hk + |u_1|)\,\mathrm{d}t, \quad k \geq 0. \tag{10.29}$$

We solve this problem in the usual way. The Hamiltonian has the form

$$H = -hk - |u_1| + z_1 x_2 + z_2 x_3 + z_3 u_1, \tag{10.30}$$

so that we choose the optimal control to satisfy the conditions

$$u_1 = h\,\mathrm{sgn}(z_3) \quad \text{when } |z_3| > 1, \quad u_1 = 0 \quad \text{when } |z_3| < 1. \tag{10.31}$$

The co-state equations are

$$\dot{z}_1 = 0, \quad \dot{z}_2 = -z_1, \quad \dot{z}_3 = -z_2, \tag{10.32}$$

with solutions of the form

$$z_1 = A, \quad z_2 = B - At, \quad z_3 = C - Bt + \tfrac{1}{2}At^2, \tag{10.33}$$

where A, B, and C are constants. The co-state variable z_3 is a quadratic function of the time and because the control must switch whenever z_3 passes through the values ± 1, the controls in the optimal solution must form part or the whole of one of the following sequences:

$$\{h, 0, h\}, \quad \{-h, 0, -h\}, \quad \{h, 0, -h, 0, h\}, \quad \{-h, 0, h, 0, -h\}. \tag{10.34}$$

With the initial conditions as stated, it is clear that we must begin with $\dot{x}_2$ negative and end with it positive. Thus the last of these four sequences of controls is the one we should employ. It is also clear that if we employ $u_1 = -h$ for t between 0 and τ, $u_1 = 0$ for $\tau < t < \frac{1}{2}t_1 - \tau$, and $u_1 = h$ for $\frac{1}{2}t - \tau < t < \frac{1}{2}t_1$, and by so doing reach the position $x_1 = \frac{1}{2}X$ at $t = \frac{1}{2}t_1$, we can reverse the sequence of controls for $\frac{1}{2}t_1 < t < t_1$ and reach the desired terminal state. The diagrams in Fig. 10.4 show the trajectory in the (x_1, x_2) plane, the co-state variable z_3, and the changes in x_3.

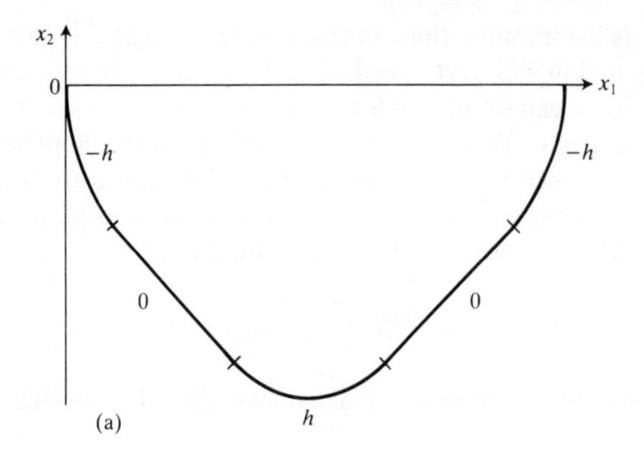

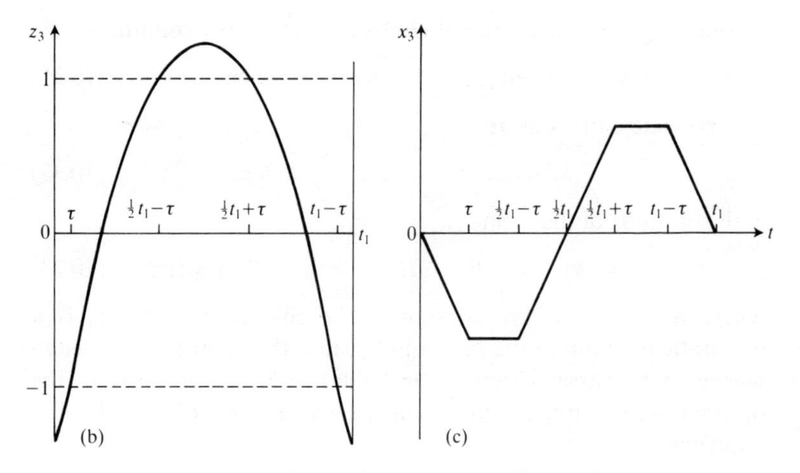

Fig. 10.4. Optimal solutions for the trolley problem.

If we solve the state equations in the three intervals from 0 to $\frac{1}{2}t_1$ and ensure that the state variables are continuous, we find that $x_1 = \frac{1}{2}X$ at $t = \frac{1}{2}t_1$ provided that

$$X = \tfrac{1}{4}ht_1\tau(t_1 - 2\tau). \tag{10.35}$$

A second equation connecting the unknowns τ and t_1 comes from consideration of z_3. We know that z_3 must have its maximum value at $t = \frac{1}{2}t_1$ because of the symmetry on either side of this time and that it must take the value $+1$ at $t = \frac{1}{2}t_1 - \tau$ and the value -1 at

$t = \tau$. This is still not enough to determine the solution, and we must use the fact that the Hamiltonian is zero along the optimal trajectory. From these conditions it follows that

$$2\tau^2 - (k + 2)\tau t_1 + \tfrac{1}{4}kt_1^2 = 0 \tag{10.36}$$

and hence $\tau = \lambda t_1$, where

$$\lambda = \tfrac{1}{2} + \tfrac{1}{4}k - \tfrac{1}{2}[1 + \tfrac{1}{2}k + \tfrac{1}{4}k^2]^{1/2}. \tag{10.37}$$

It follows from (10.35) that the optimal time is given by

$$t_1 = \left(\frac{4X}{h\lambda(1 - 2\lambda)}\right)^{1/3}, \tag{10.38}$$

and the optimal cost, from (10.29), has the value

$$J = ht_1(k + 4\lambda). \tag{10.39}$$

In the limit as $k \to \infty$, which is the time-optimal case, $\lambda = \frac{1}{4}$ and there are no intervals during which $u_1 = 0$. The optimal cost has the asymptotic value $J \sim (32h^2X)^{1/3}k$. For $k = 0$, when the cost depends only on the amount of sand used in the control, there is no optimal solution and we can make the cost arbitrarily small, although the terminal time will be arbitrarily large.

10.4. Other control sets

The controls that we have been using throughout the applications have been piecewise-continuous functions of t whose range has been either unbounded or bounded. The sets of controls have been denoted by $\mathscr{U}_{\mathrm{u}}$ when the m controls can take any value in $\mathscr{R}^m$, by $\mathscr{U}_{\mathrm{b}}$ when their range is the m-dimensional hypercube $\mathscr{V}_{\mathrm{b}}$ defined by $|u_i| \le 1$, and by $\mathscr{U}_{\mathrm{bb}}$ in the bang–bang case when their range is $\partial\mathscr{V}_{\mathrm{b}}$, the boundary of the hypercube, given by $|u_i| = 1$. The PMP, however, does not require that the control set be restricted to these sets. More generally, we can define a control set $\mathscr{U}_{\mathrm{c}}$, consisting of piecewise-continuous functions of t whose range lies in some closed convex set $\mathscr{V}_{\mathrm{c}} \subset \mathscr{R}^m$. It is not necessary that $\mathscr{V}_{\mathrm{c}}$ should be bounded; for example, we could require that $|u_1| \le 1$ while the remaining components of $\mathbf{u}$ are unbounded. An important special case is when the length of the control vector is bounded, so that $\|\mathbf{u}\| \le 1$, and $\mathscr{V}_{\mathrm{c}}$ is the closed ball in $\mathscr{R}^m$ with centre $\mathbf{0}$ and unit radius.

The PMP tells us that the optimal control maximizes the Hamiltonian over all admissible controls. This restricted maximum problem can be solved using Lagrange multipliers, but it is not usually necessary to resort to this formalism.

As an example, let us consider the time-optimal linear problem, with state equation

$$\dot{\mathbf{x}} = \mathbf{Ax} + \mathbf{Bu}. \tag{10.40}$$

The Hamiltonian is given by

$$H = -1 + \mathbf{z}^{\mathrm{T}}(\mathbf{Ax} + \mathbf{Bu}). \tag{10.41}$$

The PMP in the case leads to the condition that the optimal control is given by

$$\mathbf{z}^{\mathrm{T}}\mathbf{Bu} = \sup_{\mathbf{v} \in \mathscr{U}_c} \mathbf{z}^{\mathrm{T}}\mathbf{Bv}, \tag{10.42}$$

where $\mathbf{z}$ is the co-state variable. The geometrical interpretation of this condition is that the optimal control $\mathbf{u}(t)$ must lie on $\partial\mathscr{V}_c$ and the supporting hyperplane to $\mathscr{V}_c$ at $\mathbf{u}$ has its normal in the direction $\mathbf{B}^{\mathrm{T}}\mathbf{z}$ away from $\mathscr{V}_c$. The co-state vector varies continuously, so that if $\mathscr{V}_c$ is smooth, the optimal control will be continuous. An exception to this continuity would occur if $\mathbf{z}$ were to pass through $\mathbf{0}$, but this is impossible since the Hamiltonian would not then be zero, a contradiction of the PMP. On the other hand, if $\partial\mathscr{V}_c$ is made up of hyperplanes, as $\mathbf{z}$ varies the optimal control $\mathbf{u}$ will jump from one vertex of $\mathscr{V}_c$ to another. This is a generalization of the bang-bang behaviour of the optimal control when the control set is the hypercube $\mathscr{V}_b$. The following two examples demonstrate these principles.

1. *Triangular control set*

Suppose that there are two controls u_1 and u_2 and that their values lie inside or on the boundary of the set $\mathscr{V}_c$, the triangle with vertices at the three points A(0, −1), B(−1, 0), and C(1, 1). The two state variables are x_1 and x_2, which initially have the positive values X_1 and X_2, respectively. We wish to steer them to the origin in as short a time as possible using admissible controls, and the state equations are given by

$$\dot{x}_1 = -x_1 + u_1, \quad \dot{x}_2 = u_2. \tag{10.43}$$

The Hamiltonian is given by

$$H = -1 + z_1(-x_1 + u_1) + z_2 u_2, \tag{10.44}$$

and the co-state equations are

$$\dot{z}_1 = z_1, \quad \dot{z}_2 = 0. \tag{10.45}$$

In Fig. 10.5, the lines OD, OE, and OF are perpendicular to the sides BC, CA, and AB of the triangle $\mathscr{V}_c$. If the vector $\mathbf{z}$ lies between OE and OF, the line with normal in the direction of $\mathbf{z}$ will support $\mathscr{V}_c$ at A, and the corresponding optimal control will be given by $u_1 = 0$, $u_2 = -1$. Similarly, if $\mathbf{z}$ lies between OD and OF, the optimal controls are $u_1 = -1$, $u_2 = 0$, and if it lies between OD and OE, $u_1 = 1, u_2 = 1$.

Starting from their given positive values, and ending at the origin, the state variables must not increase, and we therefore expect that $\mathbf{z}$ will lie between OA and OB. We therefore assume that the appropriate solutions of the co-state equations (10.45) have the forms

$$z_1 = \alpha\, e^t, \quad z_2 = -\beta, \tag{10.46}$$

where α and β are positive constants. It follows that, if the controls switch, we must start from A with $\alpha < \beta$ and move to B when

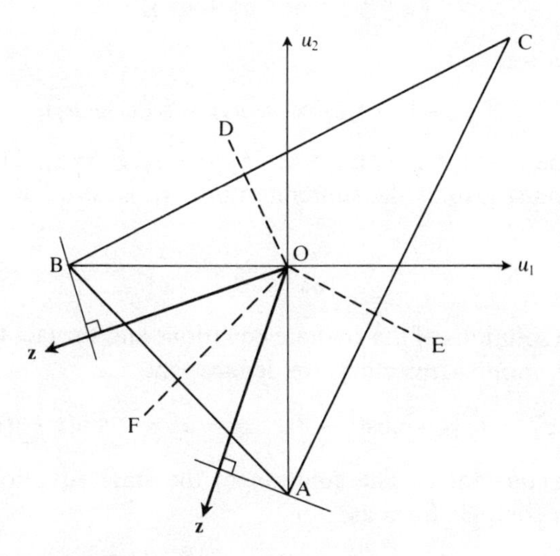

Fig. 10.5. The co-state vector and the control set.

$\alpha\, e^t = \beta$. Solving the state equations, we see that

$$x_1 = X_1 e^{-t}, \quad x_2 = X_2 - t, \quad \text{for } 0 < t < t_2, \tag{10.47}$$

and

$$x_1 = -1 + \exp(t_1 - t), \quad x_2 = 0, \quad \text{for } t_2 < t < t_1. \tag{10.48}$$

Matching the two portions of the trajectories at the switching time t_2, we find that

$$t_2 = X_2, \quad t_1 = \ln[X_1 + \exp(X_2)]. \tag{10.49}$$

For all positive values of X_1 and X_2, $t_1 > t_2 > 0$, so this is an acceptable solution. A different sequence of controls would be needed for other initial states (see Exercise 10.5).

2. *Smooth control set*

As an example of a smooth control set, we consider the harmonic oscillator with two controls. The state equations are given by

$$\dot{x}_1 = -x_2 + u_1, \quad \dot{x}_2 = x_1 + u_2, \tag{10.50}$$

with the initial state $x_1 = R\cos\phi$, $x_2 = R\sin\phi$, and the target is the origin. We wish to reach the target in minimum time with controls satisfying the condition

$$\|\mathbf{u}\| \le 1 \quad \text{or} \quad u_1^2 + u_2^2 \le 1. \tag{10.51}$$

The Hamiltonian is

$$H = -1 + z_1(-x_2 + u_1) + z_2(x_1 + u_2), \tag{10.52}$$

so that the co-state equations are $\dot{z}_1 = -z_2$, $\dot{z}_2 = z_1$. The optimal controls must give us the supremum of $z_1u_1 + z_2u_2$, and hence

$$u_1 = \frac{z_1}{\|\mathbf{z}\|}, \quad u_2 = \frac{z_2}{\|\mathbf{z}\|}. \tag{10.53}$$

From the solutions of the co-state equations and the fact that $H = 0$ along the optimal trajectory, we deduce that

$$z_1 = u_1 = -\cos(t + \theta), \quad z_2 = u_2 = -\sin(t + \theta), \tag{10.54}$$

for some constant θ. The solution of the state equations can be written in complex form as

$$x_1 + \mathrm{i}x_2 = -t\exp[\mathrm{i}(t + \theta)] + A\exp(\mathrm{i}t), \tag{10.55}$$

and from the initial and final conditions we conclude that $A = R$, $\theta = \phi$, and $t_1 = R$. The optimal trajectory is the spiral given by

$$x_1 = (R - t)\cos(t + \phi), \quad x_2 = (R - t)\sin(t + \phi). \qquad (10.56)$$

In these two simple examples we have found unique optimal controls; in the second case these controls varied continuously with time, but in the first example they switched from one pair of constant values to another. When the control set is not strictly convex we can anticipate that the optimal control may not be determined uniquely. We encountered this possibility for non-normal problems with the control set $\mathscr{U}_b$ in §4.5.

10.5. Restricted state variables

Another restriction that is sometimes required by the physical situation is that the n state variables needed to model the problem may be constrained to lie within some region of the space $\mathscr{R}^n$. A simple example occurred in the Moon-landing problem of §10.1, where the module was not allowed to penetrate the surface of the Moon; the state variable was restricted to the half-space $x_1 \geq 0$. Suppose that the state vector $\mathbf{x}$ must lie within or on the boundary of the set $\mathscr{G}$, defined by $g(\mathbf{x}) \leq 0$. If the optimal trajectory without this restriction lies entirely in $\mathscr{G}$, then the optimal solution is clearly unaffected by the restriction. It is also clear that, in general, the optimal trajectory will consist of arcs that either lie in Int $\mathscr{G}$ or in $\partial\mathscr{G}$. When the state vector moves along arcs lying in $\partial\mathscr{G}$, its rate of change must be orthogonal to the normal to $\partial\mathscr{G}$ and this extra condition in the formulation of the optimal control problem must be compensated by some additional freedom. If the state equations have their general form (6.19), we modify the Hamiltonian H by adding a extra term with an unknown Lagrange multiplier. We write H in the form

$$H = -f_0 + \mathbf{z}^{\mathrm{T}}\mathbf{f} + \lambda[\operatorname{grad} g(\mathbf{x})]^{\mathrm{T}}\mathbf{f}. \qquad (10.57)$$

For portions of the trajectory not lying in $\partial\mathscr{G}$, we have $\lambda = 0$, and the extra term also vanishes when the trajectory lies in $\partial\mathscr{G}$, since $\dot{\mathbf{x}} = \mathbf{f}$. Using this extended form of the Hamiltonian, we can proceed in the usual way to write down the co-state equations and apply the maximum principle. The form that the solution takes when the trajectory is in $\partial\mathscr{G}$ is different from that when it is in the interior and it is necessary to provide jump conditions to allow the different sections to be matched together. In order to maintain the constant zero value of H along the optimal trajectory, it is clear that the quantity $\mathbf{z} + \lambda \operatorname{grad} g$ must be continuous at the points where the trajectory enters and leaves $\partial\mathscr{G}$. There is some arbitrariness in the multiplier λ to the extent of an added constant; this enables us to impose the further restriction that $\lambda = 0$ on entry to $\partial\mathscr{G}$ so that $\mathbf{z}$ is continuous there. After leaving $\partial\mathscr{G}$ we must have $\lambda = 0$ so that, in general, there will be a jump in the value of λ and also in $\mathbf{z}$ since $\mathbf{z} + \lambda \operatorname{grad} g$ is continuous. If the optimal control $\mathbf{u}$ is continuous, $\mathbf{f}$ will be continuous at the entry and exit points on $\partial\mathscr{G}$, so that the optimal trajectory will enter and leave $\partial\mathscr{G}$ tangentially. This will not necessarily be so when the control set is not smooth, as the optimal control may well switch its value on entry and exit.

1. *The positioning problem with limited speed*

As an example of a restricted state variable, suppose we consider the familiar positioning problem with a bounded control, but let us add the restriction that the speed must not exceed a limiting value. We take the state equations and the initial and final conditions as follows:

$$\dot{x}_1 = x_2, \quad \dot{x}_2 = u_1, \quad x_1^0 = X, \quad x_2^0 = 0, \quad x_1^1 = 0, \quad x_2^1 = 0, \tag{10.58}$$

with $|u_1| \leq 1$, and we want to reach the target in the shortest possible time. We choose the speed limit to restrict x_2 to the interval $-1 \leq x_2 \leq 1$, so that $\mathscr{G}$ is the infinite strip

$$g(x_1, x_2) \equiv x_2^2 - 1 \leq 0 \tag{10.59}$$

and $\partial\mathscr{G}$ consists of the two lines $x_2 = 1$ and $x_2 = -1$. The solution of this problem without the restriction was found in §6.3(3) and, for $0 < X \leq 1$, the trajectory does not lie outside $\mathscr{G}$. Hence we need only examine the case when $X > 1$.

The solution of this problem is intuitively obvious. We first apply an extreme control to increase the speed to its maximum permitted value, then switch off the control until the object is at a position from which it can be brought to rest at the origin by a further application of an extreme control; see Fig. 10.6.

To see how the general procedure outlined above applies to this problem, we first form the extended Hamiltonian, defined by

$$H = -1 + z_1 x_2 + z_2 u_1 + 2\lambda x_2 u_1, \tag{10.60}$$

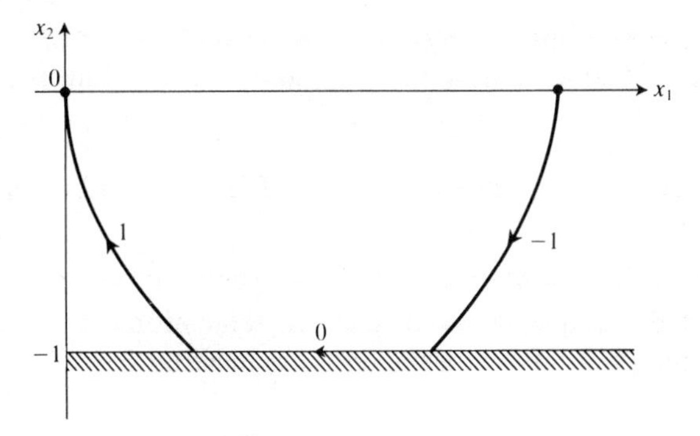

Fig. 10.6. The positioning problem with restricted speed.

and the co-state equations are

$$\dot{z}_1 = 0, \quad \dot{z}_2 = -z_1 - 2\lambda u_1. \tag{10.61}$$

When the trajectory is in Int $\mathscr{G}$, $\lambda = 0$ and the supremum of H occurs when $u_1 = \text{sgn}(z_2)$. When the trajectory is in $\partial\mathscr{G}$, that is, in the line $x_2 = -1$ in this example, we must have $u_1 = 0$ from the state equations (10.58) and this choice of control will only make H achieve a supremum if $z_2 + 2\lambda x_2 = 0$. The jump conditions are that both z_1 and $z_2 + 2\lambda x_2$ are continuous where the trajectory enters and leaves the boundary $x_2 = -1$. It follows that z_1 is continuous everywhere and that $z_2 = 0$ at the entry point.

For the first section of the trajectory, it is clear that we must have

$$\begin{aligned} &u_1 = -1, \quad x_2 = -t, \quad x_1 = X - \tfrac{1}{2}t^2, \quad z_1 = A, \\ &z_2 = -1 - At, \quad \lambda = 0, \end{aligned} \tag{10.62}$$

where we have used the condition $H = 0$ to fix the initial value of z_2. This trajectory reaches the boundary when $t = 1$ and the entry value of z_2 fixes A to be equal to -1. For $t > 1$ we have

$$u_1 = 0, \quad x_2 = -1, \quad x_1 = X + \tfrac{1}{2} - t, \quad z_2 = t - 1, \quad \lambda = \tfrac{1}{2}(t - 1). \tag{10.63}$$

Suppose the boundary of $\mathscr{G}$ is left at $t = t_2 > 1$. Then for $t_2 < t < t_2 + 1$, we must make use of an extreme control, so that

$$\begin{aligned} &u_1 = 1, \quad x_2 = t - 1 - t_2, \quad x_1 = \tfrac{1}{2}(t - 1 - t_2)^2, \\ &z_2 = t - t_2, \quad \lambda = 0. \end{aligned} \tag{10.64}$$

The jump conditions at $t = t_2$ are satisfied, and if we equate the two values of x_1 then we find that $t_2 = X$, so the optimal time is given by

$$t_1 = t_2 + 1 = X + 1. \tag{10.65}$$

For $0 < X \leq 1$ the optimal time, as found before, is given by $t_1 = 2X^{1/2}$.

2. *Positioning with limited speed and a quadratic cost*

As a second example, let us consider the same problem, but without any bounds on u_1 and with the cost J defined by

$$J = \int_0^{t_1} (2 + \tfrac{1}{2} u_1^2) \, dt. \tag{10.66}$$

When there is no restriction on x_2, the optimal solution can be found in the usual way. If $0 < X \leq \frac{4}{3}$ the optimal trajectory does not cross the line $x_2 = -1$, so we need only consider $X > \frac{4}{3}$. The optimal control and the co-state equations are now given by

$$u_1 = z_2 + 2\lambda x_2, \quad \dot{z}_1 = 0, \quad \dot{z}_2 = -z_1 - 2\lambda x_2. \tag{10.67}$$

Again the optimal trajectory is formed from three sections; in the first and third of these, $\lambda = 0$ and $x_2 > -1$, and in the second section, $x_2 = -1$ and $u_1 = 0$. The conditions at the joins between the first and second and between the second and third sections are that z_1 and $z_2 - 2\lambda$ are both continuous there. Piecing together the solutions in the three sections we find that $z_1 = -2$ throughout and that the solution has the following form:

for $0 < t < 1$

$$u_1 = z_2 = -2 + 2t, \quad \lambda = 0, \quad x_2 = t^2 - 2t, \quad x_1 = X - t^2 + \tfrac{1}{3}t^3, \tag{10.68}$$

for $1 < t < t_2 = X - \frac{1}{3}$

$$\begin{aligned} &u_1 = 0, \quad z_2 = -2 + 2\exp(t - 1), \quad \lambda = \tfrac{1}{2}z_2, \quad x_2 = -1, \\ &x_1 = X + \tfrac{1}{3} - t, \end{aligned} \tag{10.69}$$

for $t_2 < t < t_1 = X + \frac{2}{3}$,

$$\begin{aligned} &u_1 = z_2 = 2t - 2t_2, \quad \lambda = 0, \quad x_2 = -1 + (t - t_2)^2, \\ &x_1 = \tfrac{2}{3} - (t - t_2) + \tfrac{1}{3}(t - t_2)^3. \end{aligned} \tag{10.70}$$

The optimal control is continuous and the trajectory is tangential to $\partial\mathscr{G}$ at entry and exit, since $\dot{x}_2 = 0$ at $t = 1$ and at $t = t_2$. The optimal trajectory is shown in Fig. 10.7.

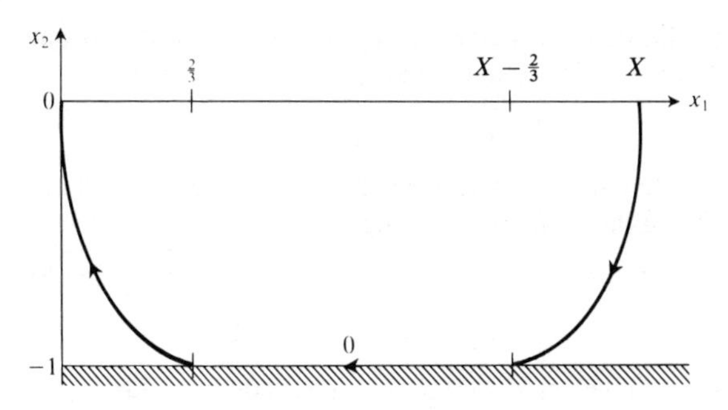

Fig. 10.7. Limited speed and a quadratic cost.

In this chapter we have discussed some more difficult applications of the PMP. The way in which more general control sets can be handled has been discussed, and we have extended the theory to allow for restrictions on the state variables. In Chapter 11 of Part C we discuss some extended optimal control problems that arise in a variety of real situations. In most practical examples the equations will be nonlinear and their solution will depend on computation. Some appropriate numerical methods for the solution of optimal control problems are described in Chapter 12.

Exercises 10

(*Solutions on p. 237*)

10.1 Solve the Moon-landing problem, as formulated in §10.1, but with a cost function that depends on the time as well as the fuel consumed. If the cost is defined by

$$J = \int_0^{t_1} (k + |u_1|)\,\mathrm{d}t, \quad k > 0,$$

and the initial state is $x_1 = h, x_2 = 0$, show that the sequence of controls in the optimal solution is $\{0, 1\}$ if $k \le c$ and $\{-1, 0, 1\}$ if $k > c$.

Find the optimal cost and the terminal time in both cases.

10.2 The state equations for the harmonic oscillator with two controls are

$$\dot{x}_1 = -x_2 + u_1, \quad \dot{x}_2 = x_1 + u_2,$$

and the cost is given by

$$J = \int_0^{t_1} \tfrac{1}{2}(1 + u_1^2 + u_2^2)\,\mathrm{d}t.$$

If the initial state is given by $x_1(0) = 0$, $x_2(0) = X > 0$, and the final state by $x_2(t_1) = 0$, show that t_1 satisfies the equation

$$t_1^2 - 2X^2 t_1 \sin t_1 \cos t_1 = X^2 \cos^2 t_1.$$

Show that there is a root of this equation between 0 and $\frac{1}{2}\pi$, and calculate the optimal cost in terms of t_1 and X.

10.3 The predator–prey equations for the problem defined in §5.3 are

$$\dot{x}_1 = x_2 + u_1, \quad \dot{x}_2 = -x_1, \quad |u_1| \le 1,$$

with initial values $x_1 = 0$, $x_2 = 1 + c$, and final values $x_1 = 0$, $x_2 = 1$. The total number of predators introduced is proportional to

$$J = \int_0^{t_1} (1 + u_1)\,dt.$$

Find solutions that minimize the value of J for an unspecified terminal time t_1 and for $0 < c \le 4$. Is there an optimal solution?

10.4 Find the optimal solution for the positioning problem with smoothly varying forces defined in §10.3 with the same conditions except that neither the initial nor the final value of x_3 is specified.

10.5 Find the control sequence for the first optimal control problem described in §10.4 for initial states that do not lie in the first quadrant.

10.6 The banks of a river are straight and lie along the lines $x_2 = 0$ and $x_2 = 1$, where x_1 and x_2 are Cartesian coordinates with x_2 measured across the river and x_1 upstream. A boat starts from the point $x_1 = X$ on the bank $x_2 = 1$ and its target is the origin. The maximum speed of the boat is equal to unity, and the river is flowing with speed V in the negative x_1 direction. Show that the state equations for the position of the boat have the form

$$\dot{x}_1 = u_1 - V, \quad \dot{x}_2 = u_2, \quad u_1^2 + u_2^2 \le 1.$$

Find the optimal solution that enables the target to be reached as quickly as possible. For what values of V does a solution exist?

Find also the solution when the objective is to cross the river without specifying the landing point.

10.7 Solve the positioning problem with $\dot{x}_1 = x_2$, $\dot{x}_2 = u_1$, and the cost function

$$J = \int_0^{t_1} \tfrac{1}{2}(k^2 + u_1^2)\,dt.$$

The initial and final positions are given by $x_1^0 = X$, $x_2^0 = 0$, $x_1^1 = 0$, $x_2^1 = 0$, and the state variables are restricted to lie in the half-space

$x_2 \geq -1$. Show that the restriction on the state variables is not active unless $X > 8/3k$. Find the terminal time and the optimal cost for all values of $X > 0$.

10.8 Show that the problem of finding the shortest distance between two points in a plane can be formulated as an optimal control problem as follows:

$$\dot{x}_1 = u_1, \quad \dot{x}_2 = u_2, \quad x_1(0) = X_1, \quad x_1(t_1) = Y_1,$$
$$x_2(0) = X_2, \quad x_2(t_1) = Y_2,$$

and the optimal control minimizes t_1 subject to the constraint $u_1^2 + u_2^2 \leq 1$. The shortest distance is equal to the optimal value of t_1. Verify by using the PMP that the optimal trajectory is the straight line joining the initial and final points.

Suppose that there is an obstacle in the plane that restricts the state variables to lie outside or on its boundary. Consider the special case when $X_1 = -2$, $Y_1 = 2$, $X_2 = 0$, $Y_2 = 0$, and the state variable is restricted by the condition $1 - x_1^2 - x_2^2 \leq 0$. Apply the methods of §10.5 to solve this problem. Show the the optimal trajectory is tangential to the boundary of the obstacle on entry and exit.

PART C
Applications of Optimal Control Theory

In Part C we first discuss some examples of problems to which optimal control theory has been successfully applied, taken from a variety of fields. We then consider briefly some numerical methods for the solution of nonlinear problems, for which analytical solutions are not available, and apply these methods to some specific problems.

11 Some applied optimal control problems

> The bearings of this observation lays in the application on it.
>
> Charles Dickens

Optimal control theory has been applied to a large variety of problems in many different fields. Some examples have been given in the preceding chapters. In this chapter we consider some applications of the theory in a more extended form, although the mathematical models are still greatly simplified. Some features of the real situations have to be ignored in order to present solutions that are meaningful to a non-specialist and that demonstrate the power of the theory in providing optimal strategies. Four problems, related to fishing, epidemics, trading, and physiology, are discussed in this chapter.

11.1. Maximizing the fish harvest

Although the PMP holds for nonlinear problems as well as for linear ones, all the examples that we have so far considered have been linear. (Exercise 7.6 has an element of nonlinearity in it, but the equations that have to be solved are linear.) This is because, in general, the solution of the nonlinear problems can only be completed by means of numerical calculation. This poses no severe obstacle to the determiniation of the solution, but it destroys the simplicity of the specific results that we have been able to present. Some of the examples considered in this chapter are nonlinear but of such a form that analytical (as contrasted with numerical) solutions can be found.

Suppose that the stock of a species of fish increases at a constant rate when they are not harvested. Fishing regulations set a maximum permitted rate for the harvesting of the fish, which is proportional to the size of the stock. Model equations expressing this system have the form

$$\dot{x} = x - ux, \quad 0 \leq u \leq h, \tag{11.1}$$

where $x(t)$ is the size of the stock of fish, ux is the harvesting rate, and h the maximum proportional rate. Suppose that initially $x = 1$ and that the terminal time is equal to T and is fixed. We wish to determine how to maximize the total number of fish caught, so we take as the cost function that we want to minimize

$$J = -\int_0^T ux \, \mathrm{d}t. \tag{11.2}$$

It is clear that we shall have different strategies depending on whether h is less than or greater than one.

Constructing the Hamiltonian in the usual way, we find that

$$H = ux + zx(1 - u), \tag{11.3}$$

so that the supremum of H is achieved if we choose $u = 0$ when $z > 1$ and $u = h$ when $z < 1$. The equation for z is given by

$$\dot{z} = -u - z(1 - u), \tag{11.4}$$

and since the target $x(T)$ is free, the transversality condition requires that $z(T) = 0$. The solutions of (11.4) have the form

$$z = -\frac{h}{1-h} + A \exp[(h-1)t] \quad \text{when } u = h, \tag{11.5}$$

and

$$z = B\,\mathrm{e}^{-t} \quad \text{when } u = 0. \tag{11.6}$$

Neither of these solutions passes through the value of $z = 1$ more than once, and it is clear from the terminal condition that we cannot end with $u = 0$, so we must either have $u = h$ for all t, or begin with $u = 0$ and switch at some time τ to $u = h$. For all t, or for $t > \tau$, the solution (11.5) for the co-state variable, with A chosen so that the transversality condition is satisfied, is given by

$$z = \frac{h}{1-h}\{\exp[(h-1)(t-T)] - 1\}. \tag{11.7}$$

We see from (11.7) that $z = 1$ when $t = \tau$, where

$$\tau = T - \delta, \quad \text{and } \delta = \frac{\ln h}{h-1}. \tag{11.8}$$

Note that δ is positive for all positive values of h; for $h = 1$ we define $\delta = 1$ so that δ is a continuous function of h. We can identify two cases, depending on the sign of τ. Suppose first that $T < \delta$ so that τ is negative. Then $u = h$ for all t; the solution of the state equation (11.1) is

$$x = \exp[(1 - h)t] \tag{11.9}$$

and the maximum catch is, from (11.2),

$$-J = \frac{h}{1 - h}\{\exp[(1 - h)T] - 1\}. \tag{11.10}$$

Secondly, if $T > \delta$, τ is positive, and we must begin by setting $u = 0$, switching to $u = h$ for $t > \tau$. The optimal solution is then given by

$$x = \mathrm{e}^t \text{ for } t < \tau, \quad x = \exp[h\tau - (h - 1)t] \quad \text{for } \tau < t < T, \tag{11.11}$$

and the optimal catch is

$$-J = h\int_\tau^T x\,\mathrm{d}t = \frac{h}{h - 1}\{\mathrm{e}^\tau - \exp[h\tau - (h - 1)T]\}. \tag{11.12}$$

When we substitute the value of τ from (11.8), and simplify, we find that the optimal catch is given by

$$-J = h^{1/(1-h)}\mathrm{e}^T. \tag{11.13}$$

The optimal catch when $h = 1$ can be found by taking the limits of these expressions as $h \to 1$, and we find that

$$-J = T \quad \text{when } T \leq 1, \quad -J = \exp(T - 1) \quad \text{when } T > 1. \tag{11.14}$$

In the optimal solution, the time during which fishing takes place is equal to δ when T is greater than δ and is then independent of T. Because the fishing rate is proportional to the stock, it is advantageous to allow the stock its natural increase for as long as possible.

Although the state equation in this example is nonlinear, the controls to be used are bang–bang, so that the nonlinearity does not introduce any difficulty into the solution. The model for the growth of the stock is of limited validity, since there is usually a natural limit to the size of the stock that can be maintained by its environment. A better model includes a quadratic term in the equation for the growth of the stock, which also introduces an additional nonlinear term. We replace (11.1) by the equation

$$\dot{x} = 2x - x^2 - ux, \tag{11.15}$$

so that, without depletion, the stock would equilibrate at the value $x = 2$. Suppose we start with $x(0) = 2$, with the other conditions of the problem the same as before. The Hamiltonian now has the form

$$H = xu + z(2x - x^2 - xu), \tag{11.16}$$

and the co-state equation is

$$\dot{z} = -u - z(2 - 2x - u). \tag{11.17}$$

To find the supremum of H we maximize $xu(1 - z)$, and, since x is positive, we find that, as before, $u = h$ when $z < 1$ and $u = 0$ when $z > 1$. Since we are starting from the equilibrium state, there is no point in commencing with $u = 0$, so that we may take $z(0) < 1$. Since H is constant along the trajectory, we can sketch the phase curves in the (x, z) plane for $u = 0$ and for $u = h$, which have equations of the form

$$z = \frac{c}{2x - x^2}, \quad z = \frac{c - xh}{2x - x^2 - xh}, \tag{11.18}$$

respectively, where c is a constant. The curves with $u = 0$ have their minima at $x = 1$, and those with $u = h$ have maxima along the curve

$$z = \frac{-h}{2 - 2x - h}, \quad \text{for } x > 1 - \tfrac{1}{2}h. \tag{11.19}$$

Some curves for the case when $1 < h < 2$ are shown in Fig. 11.1. The curves starting from $x = 2$ with $z < 1$ cross the x axis at a point to the right of $x = 2 - h$ and must reach this point at time T, since

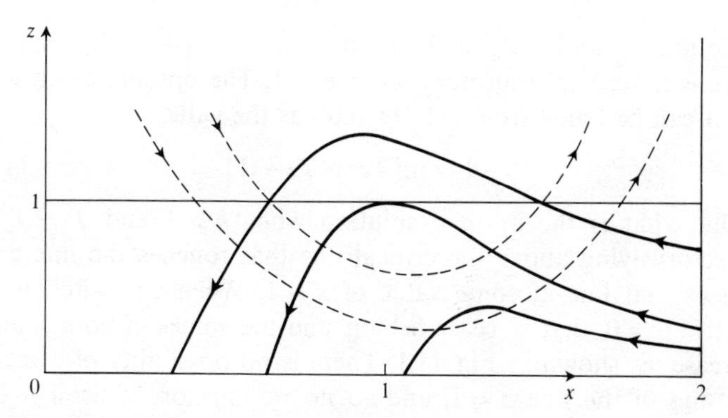

Fig. 11.1. The solid lines are trajectories with $u = h$ and the broken lines with $u = 0$.

$z(T) = 0$. However, $z(0)$ must be sufficiently small to ensure that the trajectory does not cross the line $z = 1$. Since the locus of maxima crosses this line at $x = 1$, the critical trajectory is the one that passes through the point (1, 1).

Because the trajectories do not cross the line $x = 2 - h$, there is no critical trajectory when $h < 1$. For all solutions when $h < 1$, and for the subcritical ones when $1 < h < 2$, we have $u = h$ for all t and the solutions of the state and co-state equations have the form

$$x = \frac{2(2-h)\exp[(2-h)t]}{2\exp[(2-h)t] - h}, \quad z = \frac{h[x(T) - x]}{2x - x^2 - xh}. \tag{11.20}$$

The optimal catch is given by

$$-J = h\int_0^T x\,\mathrm{d}t = h\ln\left(\frac{2\exp[(2-h)T] - h}{2-h}\right). \tag{11.21}$$

This is the solution for all T when $h < 1$, and for $T \leq T_c$ when $1 < h < 2$, where T_c is the value of T for which the trajectory goes through (1, 1). The value of $x(T_c)$ can be found from the second equation in (11.20) and when this value is substituted in the first equation in (11.20) we obtain an equation for T_c, which has the solution

$$T_c = \frac{1}{2-h}\ln\left(\frac{h}{2(h-1)^2}\right). \tag{11.22}$$

The optimal catch in this critical case has the value

$$-J_c = 2h\ln\left(\frac{h}{h-1}\right). \tag{11.23}$$

Note that T_c and $-J_c$ tend to infinity as $h \to 1+$, indicating that there is no critical trajectory when $h = 1$. The optimal catch when $h = 1$ can be found from (11.21) and has the value

$$-J = \ln[2\exp(T) - 1]. \tag{11.24}$$

But what is the optimal solution when $h > 1$ and $T > T_c$? A trajectory lying above the critical one that touches the line $z = 1$ crosses that line at some value of $x > 1$. We must switch to the control $u = 0$, that is, cease fishing, and the values of both x and z increase, as shown in Fig. 11.1. There is no possibility of a second crossing of the line $z = 1$, and so no resumption of fishing. It is impossible to satisfy the terminal condition $z(T) = 0$, so this cannot be the optimal solution. To find a way through this impasse, let us look again at the supremum of H, given in (11.16). We deduced that we must use either $u = h$ or $u = 0$, switching from one to the other when z passed through the value 1. But suppose that $z = 1$ for an interval of time and not only instantaneously. Then any value of u will give us the supremum of H. But if z remains at the value 1, we must have $\dot{z} = 0$ and, from (11.17), this can only occur at $x = 1$. Thus we must also have $\dot{x} = 0$, and so, from (11.15), $u = 1$. This means that provided we can reach the point (1, 1) we can stay there indefinitely and fish at the rate $u = 1$. The fishing rate is exactly equal to the rate at which the stock would grow naturally at this stock size, so it can be maintained. To complete the optimal solution we must end with $z(T) = 0$. To do this we follow the critical trajectory from (1, 1) with $u = h$ until $z = 0$. The optimal strategy is, therefore, to fish at the maximum rate h for a time equal to τ_1, say, when $z = 1$. We then reduce the fishing rate to one, and continue for a time τ_2. Finally, we revert to the maximum rate h for a time τ_3. The trajectory in Fig. 11.1 is the same as the critical one, except that we pause at the point (1, 1) for time τ_2, so that $\tau_1 + \tau_3 = T_c$ and $\tau_2 = T - T_c$. The optimal catch is given by

$$
\begin{aligned}
-J &= -J_c + \int_{\tau_1}^{\tau_1+\tau_2} 1\,\mathrm{d}T = -J_c + T - T_c \\
&= T + 2h\ln\left(\frac{h}{h-1}\right) - \frac{1}{2-h}\ln\left(\frac{h}{2(h-1)^2}\right). \qquad (11.25)
\end{aligned}
$$

The limit of this expression as $h \to 1+$ is $T + \ln 2$, but $T_c \to \infty$. For $h = 1$ the optimal catch is given by (11.24), which, for large T, becomes

$$-J = T + \ln 2 - \tfrac{1}{2}\exp(-T) + O(\exp(-2T)). \qquad (11.26)$$

When $h \to 2$, the optimal catch, from (11.25), is $T + 4\ln 2 - \frac{3}{2}$. The analysis when $h > 2$ follows the same lines as that for h between 1 and 2, and (11.25) is still valid. In the limit as $h \to \infty$ the optimal catch is $T + 2$.

To summarize these results, we have found that, for large T, the optimal strategy when $h \leq 1$ is to fish at the maximum rate for all t, the optimal catch being given approximately by

$$-J = h(2-h)T + h\ln\left(\frac{2}{2-h}\right) - \tfrac{1}{2}h^2\exp[-(2-h)T]. \qquad (11.27)$$

When $h > 1$, the strategy is to fish at the maximum rate until the stock is reduced to half the equilibrium size and then to fish at the rate $u = 1$ for most of the time. This is followed by a short period during which fishing at the rate $u = h$ is resumed. The optimal catch is then given by (11.25). Thus, increasing the maximum permitted rate from one to infinity only increases the optimal catch from $T + \ln 2$ to $T + 2$, and very little is to be gained by fishing at a rate in excess of $u = 1$. Only if h is less than one should the fishing be at the maximum permitted rate.

11.2. Control of epidemics

In the fishing problem of the last section, the state and co-state equations were nonlinear but, because there was only a single state variable and the optimal control was bang–bang, we were able to find an analytical expression for the solution without having to resort to numerical evaluation. Nonlinear systems of higher dimension are unlikely to be so tractable. The next problem cannot be solved except with the aid of numerical methods for solving the differential equations.

A simplified model for the spread of an epidemic divides the population into four classes: (i) those who are susceptible to the disease but are not yet infected, (ii) those who have the disease, (iii) those who die from it, (iv) those who are unaffected by the disease. This last category includes those who are naturally immune, those who survive the infection and become immune, and those who are isolated from the centres of infection. The natural process for an individual is to move from class (i) through class (ii) and into either class (iii) or (iv). With a programme of inoculation it is possible to ensure that some individuals can pass straight from class (i) to class (iv). We can ignore the members of class (iv), and suppose that the proportions of the remaining population in each of the other classes are measured by x_1, x_2, and x_3, respectively. Initially, we may suppose that the proportion of individuals that are infected is given by $x_2(0) = \delta > 0$, with $x_1(0) = 1 - \delta$ and $x_3(0) = 0$. The large number of potential sufferers from the disease leads to its spread throughout the population. The objective is to reduce the infected class to a low positive value $\varepsilon < \delta$ after an unspecified time interval during which the number of uninfected people declines. We can assume that a constant fraction of those infected die from the disease, the others being transferred to class (iv). The cost is measured by the numbers of those who die and by the cost of the inoculation programme. A set of state equations that model the system can be taken in the following form:

$$\dot{x}_1 = -rx_2x_1 - u, \tag{11.28}$$

$$\dot{x}_2 = rx_1x_2 - sx_2, \tag{11.29}$$

$$\dot{x}_3 = csx_2, \tag{11.30}$$

$$J = \int_0^{t_1} bu^2 \, \mathrm{d}t + (1-b)x_3(t_1). \tag{11.31}$$

In these equations, the nonlinear terms with positive coefficient r represent the fact that the rate of growth in the numbers infected is proportional to the uninfected population, the term with positive coefficient s gives the rate at which the number of those infected is reduced by recovery and death, and c is the proportion of those infected who die. When the pool of uninfected persons becomes empty it remains empty and the variable x_1 must satisfy the restriction $x_1 \geq 0$. Such restrictions on state variables were discussed in §10.5. The rate of inoculation is given by u and the cost function contains a component that increases quadratically with u, indicating the increased cost of large-scale inoculations. The other component in the cost is the total number of those who die and the relative weighting of the two components is given by the constant b, where $0 \leq b \leq 1$.

It is clear that we can eliminate x_3 from consideration by combining (11.30) and (11.31) to give the cost function in the form

$$J = \int_0^{t_1} [bu^2 + (1-b)csx_2] \, \mathrm{d}t. \tag{11.32}$$

The initial values have already been fixed. In order that x_2 should increase initially we must have $r(1-\delta) - s > 0$; otherwise the disease will not spread to any significant extent. The terminal time $t_1 > 0$ is free and there are two possibilities for the terminal state. We must reach the state $x_2(t_1) = \varepsilon$, but $x_1(t_1)$ may be positive or zero. If positive, the transversality condition shows that the co-state variable $z_1(t_1) = 0$. The other possibility is that $x_1(t_1) = 0$ and that a portion of the trajectory lies on this boundary of the permitted region, as in §10.5. The reason for the requirement that x_2 should reach the value

ε and not zero is that x_2 decreases exponentially when $rx_1 < s$, so that it would take an infinite time for x_2 to be reduced to zero. It is easy to find the finite additional cost incurred as the epidemic tails off.

The Hamiltonian for this system can be found in the usual way, with the two co-state variables z_1 and z_2 introduced. The supremum of H is attained if we choose $u = -\frac{1}{2}z_1/b$ when $x_1 > 0$ and $u = 0$ when $x_1 = 0$. In both cases, the state and co-state equations and the cost function have the forms

$$\dot{x}_1 = -rx_2x_1 + z_1/2b, \tag{11.33}$$

$$\dot{x}_2 = rx_1x_2 - sx_2, \tag{11.34}$$

$$\dot{z}_1 = -rz_2x_2 + rx_2z_1, \tag{11.35}$$

$$\dot{z}_2 = (1 - b)cs + rz_1x_1 - (rx_1 - s)z_2, \tag{11.36}$$

$$J = \int_0^{t_1} \left(\frac{z_1^2}{4b} + (1 - b)csx_2\right) \mathrm{d}t. \tag{11.37}$$

The initial conditions are

$$x_1(0) = 1 - \delta, \quad x_2(0) = \delta, \tag{11.38}$$

and the final conditions are, when $x_1(t_1) > 0$,

$$x_2(t_1) = \varepsilon, \quad z_1(t_1) = 0, \tag{11.39}$$

and otherwise,

$$x_2(t_1) = \varepsilon, \quad x_1(t_1) = 0. \tag{11.40}$$

Finally, we have the condition that $H = 0$ along the optimal trajectory and, applying this condition at $t = t_1$, we find that

$$-cs(1 - b) + z_2(rx_1 - s) = 0. \tag{11.41}$$

For large values of b we expect a low inoculation rate and consequently that the number infected drops to the desired level before the pool of uninfected persons is exhausted, so that the conditions (11.39) hold. When b is small, however, we can employ a large rate of inoculation which eliminates the susceptible population very quickly, and the conditions (11.40) are appropriate. If we suppose that x_1 becomes zero at $t = t_2$, from t_2 to t_1 the solutions of (11.33), (11.34), and (11.36) are of the form

$$x_1 = 0, \quad x_2 = X_2 \exp[-s(t - t_2)], \quad z_2 = -c(1 - b), \tag{11.42}$$

where $x_2(t_2) = X_2$. If we evaluate the Hamiltonian as $t \to t_2-$, we find that $z_1(t_2) = 0$, and, since x_1 remains zero, (11.33) then shows that $z_1(t) = 0$ for $t > t_2$. It follows that we have to solve the four state and co-state equations for $0 < t < t_2$ with $x_1 > 0$ and with the conditions

$$x_1(t_2) = 0, \quad x_2(t_2) = X_2, \quad z_1(t_2) = 0, \quad z_2(t_2) = -c(1-b). \quad (11.43)$$

The solution that satisfies these conditions and the two initial conditions (11.38) also fixes the values of X_2 and t_2. The value of t_1 follows from (11.42), since $x_2(t_1) = \varepsilon$, and so

$$t_1 = t_2 + \frac{1}{s}\ln\left(\frac{X_2}{\varepsilon}\right). \quad (11.44)$$

If J_2 is the value of the cost integral over the interval from 0 to t_2, the optimal cost is given by

$$J = J_2 + (1-b)cs\int_{t_2}^{t_1} x_2 \, dt = J_2 + (1-b)c(X_2 - \varepsilon).$$

For general values of b, of course, we do not know which of the alternative end conditions we should employ until the calculation is under way.

For all values of b, the equations that have to be solved to determine the optimal control and the corresponding cost are nonlinear. We also have a boundary-value problem, with two conditions at each end of the time interval. Moreover, the terminal time has itself to be determined as part of the solution, and the terminal conditions have alternative forms which are solution dependent. This all adds up to a difficult numerical task, and suitable methods for its accomplishment are discussed in Chapter 12. The simplest way of overcoming these difficulties is to reverse the time direction. The four conditions at $t = t_2$ are known if we specify the value of X_2. Integrating this autonomous initial-value problem for t decreasing, we can continue until the solution satisfies the condition $x_1 + x_2 = 1$. The time interval from the start of the calculation fixes t_2 and the final value of x_2 fixes δ. Similarly, when the final conditions are given by (11.39), we can assume a value X_1 for $x_1(t_1)$, and the values of x_2, z_1, and z_2 at this time are fixed by (11.39) and (11.41). Integrating backwards in time, as before, we continue until $x_1 + x_2 = 1$ and t_1 and δ are then determined. By varying the values of X_1 and X_2 we can find a set of solutions for various values of δ, that is, for different initial levels of infection. The solution for a prescribed value of δ can be found by interpolation. The results of a numerical solution obtained by this procedure are shown in Fig. 11.2 for the parameter values $r = 1.0$, $s = 0.4$, $b = 0.5$, $c = 0.5$, $\delta = 0.1$, $\varepsilon = 0.01$. The diagram shows the growth and decline in the number of persons infected in the uncontrolled case, with $u = 0$, and in the optimal solution. In the optimal solution x_1 becomes zero at $t = 9.0$, when $x_2 = 0.02$, and the terminal time is $t_1 = 10.73$; without inoculation, the terminal time is $t_1 = 17.6$. The optimal cost is $J = 0.129$; the cost measured by the proportion of deaths is 0.179 with inoculation and 0.447 without.

In dealing with nonlinear problems, it is often possible to find limiting forms for their solution when some of the parameters have extreme values, such as zero or infinity. Techniques for obtaining such solutions are well known in various fields of applied mathematics; they involve the identification of different scales on which the variables change their values. The most widely known example is the concept of the boundary layer in fluid mechanics. We can apply such a method to the present problem if we consider the solution when the parameter b is very small.

If we were to set $b = 0$, there would be no limit on the size of the inoculation rate u and we could eliminate the susceptible population

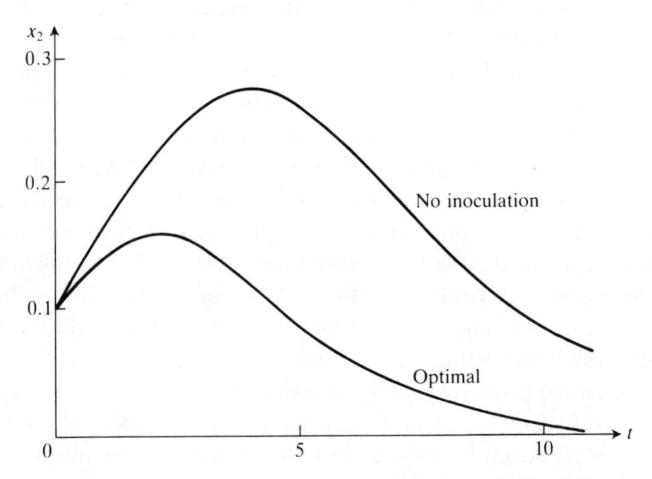

Fig. 11.2. Progress of an epidemic with no inoculation and with the optimal control.

x_1 immediately. The value of x_2 does not change while this rapid change in x_1 is progressing. Hence, the effective initial values become $x_1 = 0$ and $x_2 = \delta$. The size of the infected population then decreases exponentially, as there is no supply of hitherto uninfected persons to add to it, and the optimal solution, optimal time, and terminal time are given by

$$x_2 = \delta \exp(-st), \quad J = c(\delta - \varepsilon), \quad t_1 = \frac{1}{s}\ln\left(\frac{\delta}{\varepsilon}\right). \tag{11.45}$$

In this solution, which is the asymptotic form of the solution as $b \to 0$, we can identify two different timescales, a rapid one in which x_1 changes and a slower one during which x_2 decreases. Different terms in the governing equations are important in the two regimes. We refer to the first stage, in which there is a rapid variation for a short time, as the inner region and the second stage as the outer region. The solution in the outer region has already been found in the discussion of the solution for small values of b, so that we can concentrate on the inner region. Because the state equations have a discontinuity in their form at $x_1 = 0$, we can match the two regions at $t = t_2$ without having to relate the limiting forms in the inner and outer regions as is usually required in problems of this type. To find the solution in the inner region, let us write $b = \theta^2$, where $0 < \theta \ll 1$, and introduce scaled variables τ and ζ_1, where

$$t = \theta\tau, \quad t_2 = \theta\tau_2, \quad z_1 = \theta\zeta_1. \tag{11.46}$$

The state and co-state equations (11.33)–(11.36) in the new variables have the forms

$$\frac{dx_1}{d\tau} = \tfrac{1}{2}\zeta_1 - \theta r x_1 x_2, \tag{11.47}$$

$$\frac{dx_2}{d\tau} = \theta x_2(rx_1 - s), \tag{11.48}$$

$$\frac{d\zeta_1}{d\tau} = -rx_2 z_2 + \theta r x_2 \zeta_1, \tag{11.49}$$

$$\frac{dz_2}{d\tau} = \theta(1 - \theta^2)cs + \theta^2 r x_1 \zeta_1 - \theta z_2(rx_1 - s). \tag{11.50}$$

The initial and final conditions in the new variables can be found from (11.38) and (11.39) and are given by

$$x_1 = 1 - \delta, \quad x_2 = \delta, \quad \text{at } \tau = 0, \tag{11.51}$$

$$x_1 = 0, \quad x_2 = X_2, \quad \zeta_1 = 0, \quad z_2 = -c(1 - \theta^2), \quad \text{at } \tau = \tau_2, \tag{11.52}$$

and, provided $X_2 > \varepsilon$, the optimal cost, from (11.37), is

$$J = c(1 - \theta^2)(X_2 - \varepsilon) + \theta \int_0^{\tau_1} [\tfrac{1}{4}\zeta_1^2 + (1 - \theta^2)csx_2] \, d\tau. \tag{11.53}$$

We now expand the variables in powers of θ, writing, for example, $x_1 = x_{10} + \theta x_{11} + \cdots$, with similar expressions for the other dependent variables and for τ_2, X_2, and J. The first-order equations are

$$\dot{x}_{10} = \tfrac{1}{2}\zeta_{10}, \quad \dot{x}_{20} = 0, \quad \dot{\zeta}_{10} = -rx_{20}z_{20}, \quad \dot{z}_{20} = 0, \tag{11.54}$$

and their solutions that satisfy the end conditions are given by

$$z_{20} = -c, \quad x_{20} = \delta, \quad \zeta_{10} = \delta cr(\tau - \tau_{20}), \tag{11.55}$$

$$x_{10} = 1 - \delta - \tfrac{1}{4}\delta cr\tau_{20}^2 + \tfrac{1}{4}\delta cr(\tau - \tau_{20})^2. \tag{11.56}$$

Since we must also satisfy the condition $x_{10}(\tau_{20}) = 0$, the value of τ_{20} is given by

$$\tau_{20} = 2\left(\frac{1 - \delta}{\delta cr}\right)^{1/2}, \tag{11.57}$$

and, since $X_{20} = \delta$, the leading term in the optimal cost is given by

$$J_0 = c(\delta - \varepsilon). \tag{11.58}$$

The optimal control, in the original variables, is given by

$$u = \frac{\delta cr}{2\theta^2}(\theta\tau_{20} - t), \quad \text{for } 0 < t < \theta\tau_{20}, \tag{11.59}$$

so that the inoculation rate is high, but decreases linearly with time until the susceptible class is empty.

To find the solution to order θ, we note that part of the contribution to J at this order only involves the leading-order terms ζ_{10} and x_{20}, but we also need to know the value of X_{21}. Since x_{20} is a constant, $x_2(\tau_2) = x_{20}(\tau_{20}) + \theta x_{21}(\tau_{20}) + \cdots$. The equation for x_{21} is, from (11.48),

$$\dot{x}_{21} = x_{20}(rx_{10} - s) \tag{11.60}$$

and the solution of this equation has the form

$$x_{21} = \tfrac{1}{12}\delta^2 cr^2[(\tau - \tau_{20})^3 + \tau_{20}^3] - \delta s\tau. \qquad (11.61)$$

Hence we find that

$$X_{21} = 2\delta\left(\frac{1-\delta}{\delta cr}\right)^{1/2}[\tfrac{1}{3}r(1-\delta) - s]. \qquad (11.62)$$

We can now find the optimal cost to this order by inserting the values we have determined into (11.53); after some simplification we obtain the result

$$J = c(\delta - \varepsilon) + \tfrac{4}{3}\theta[\delta cr(1-\delta)^3]^{1/2}. \qquad (11.63)$$

To complete the solution at this order, we must solve the equations for x_{11}, ζ_{11}, and z_{21} and determine the value of τ_{21}, which is a straightforward exercise.

11.3. Commodity trading

The next application is to a simple economic system. The buying and selling of a commodity, such as wheat, coffee, or oil, by traders who do not intend to make use of the commodity themselves is intended by those who engage in this traffic to yield a profit. The possibility of success depends on the seasonal or longer-term fluctuations of the price of the commodity, and a large part of the skill of a successful trader depends on the ability to make an accurate forecast of the future price for as long a period as possible. Given a forecast for the behaviour of the price, it is possible to pose an optimal control problem to determine when the commodity should be bought or sold and when the trader should be inactive. Suppose a forecast of the price is made for a certain period T, which is as far as one can see ahead and is known as the planning horizon. Then the optimal control problem is to maximize the assets at time T, these assets consisting of the total value of the cash and the commodity held then. Afterwards, when the actual price behaviour is a matter of record, it is possible to calculate what would have been the maximum profit if this behaviour had been known and the corresponding optimal strategy employed; a comparison between the realized and optimal profit could be used as a performance indicator to measure the trader's ability. In practice the operations of buying and selling will be discrete but here we employ a continuous model which is both informative and easy to use.

Suppose the amounts of cash and the commodity held at time t are denoted by $x_1(t)$ and $x_2(t)$, respectively. The price at time t is the known function $p(t)$ and the selling rate $u_1(t)$ lies in the interval $-1 \le u_1 \le 1$, negative values of u_1 corresponding to a buying phase of the operation. The holding of the commodity incurs a storage charge proportional to the amount of the commodity held. The state equations have the form

$$\dot{x}_1 = p(t)u_1(t) - sx_2(t), \quad \dot{x}_2 = -u_1(t), \tag{11.64}$$

where s is the positive constant of proportionality in the storage charge. Suppose we begin at $t = 0$ with a zero holding of the commodity and with an amount X_1 of cash. The final state is not specified, but the goal is to maximize the assets at $t = T$, that is, we wish to minimize the cost J, where

$$-J = x_1(T) + p(T)x_2(T). \tag{11.65}$$

The maximum profit over the period under consideration is given by

$$P = x_1(T) + p(T)x_2(T) - x_1(0) - p(0)x_2(0). \tag{11.66}$$

This problem is in the standard form and we can apply the PMP. It is a fixed-time, free-target, linear problem, with a terminal cost only. The Hamiltonian has the form

$$H = z_1(pu_1 - sx_2) - z_2u_1, \tag{11.67}$$

so that to maximize H we must choose the control specified by

$$u_1 = \operatorname{sgn}(pz_1 - z_2). \tag{11.68}$$

The co-state equations are

$$\dot{z}_1 = 0, \quad \dot{z}_2 = sz_1. \tag{11.69}$$

The initial values of the state variables are given, and the transversality condition for the free-target problem, as derived in §7.2, gives the final values of the co-state variables. These conditions are

$$x_1(0) = X_1, \quad x_2(0) = 0, \quad z_1(T) = 1, \quad z_2(T) = p(T). \tag{11.70}$$

The co-state equations (11.69) can be solved completely and we have

$$z_1 = 1, \quad z_2 = p(T) + s(t - T), \tag{11.71}$$

where we have made use of the terminal conditions. Hence the optimal control is given by

$$u_1 = \text{sgn}[p(t) - p(T) + s(T - t)], \tag{11.72}$$

and we can determine the optimal solution for any given price structure. For example, suppose that

$$\begin{aligned} p &= p_0 - p_1 t \quad \text{for } 0 < t < \tfrac{1}{2}T, \\ p &= p_0 + p_1(t - T) \quad \text{for } \tfrac{1}{2}T < t < T. \end{aligned} \tag{11.73}$$

Then the optimal solution is given by

$$\begin{aligned} u_1 &= +1, \quad x_2 = -t, \quad \text{for } 0 < t < t_s, \\ u_1 &= -1, \quad x_2 = t - 2t_s, \quad \text{for } t_s < t < T, \end{aligned} \tag{11.74}$$

where the switching time t_s is given by

$$t_s = \frac{sT}{s + p_1} < \tfrac{1}{2}T, \tag{11.75}$$

provided $s < p_1$. If $s > p_1$, the optimal solution is given by $u_1 = 1$ for all t. With $s < p_1$, we can complete the solution by determining x_1 from the first state equation in (11.64) and hence calculate the optimal profit P, defined by (11.66). Routine calculations produce the expression

$$P = \frac{p_1^2 - sp_1 + 2s^2}{4(s + p_1)} T^2. \tag{11.76}$$

In this model of the trading procedure, we have not restricted the amount of the commodity held to be non-negative. We have allowed for the possibility of shortselling, that is, the selling of goods that are not actually in the trader's possession, in the hope that by the time that settlements have to be made there will be goods to match. The solution outlined above does, in fact, involve shortselling and it is only when $t > 2t_s$ that the trader actually posseses any of the commodity. Furthermore, the model equations are not valid when there is shortselling, since then the negative value of x_2 implies that the storage charge produces a profit. If we adjust the model to forbid shortselling, we have a problem of the kind described in §10.5, in which a restriction is placed on the state variables. With the restriction $-x_2 \leq 0$ the modified Hamiltonian is

$$H = z_1(p_1 u_1 - sx_2) - z_2 u_1 + \lambda u_1. \tag{11.77}$$

The co-state equations are as before, and the jump conditions on entry or exit show that z_1 and $z_2 - \lambda$ are continuous, with $\lambda = 0$ on exit, without loss of generality. Thus $z_1 = 1$ for all t, and z_2 is a piecewise-linear function of t with slope s and $z(T) = p(T)$. When $x_2 > 0$, the solution is as before; when $x_2 = 0$, the control $u_1 = 0$. Hence the optimal solution is given by

$$\begin{aligned} u_1 &= 0, \quad x_2 = 0, \quad \text{for } 0 < t < t_s, \\ u_1 &= -1, \quad x_2 = t - t_s, \quad \text{for } t_s < t < T. \end{aligned} \tag{11.78}$$

This means that no trading is done until the time t_s has elapsed, after which the commodity is bought at the maximum rate. The optimal profit in this case is given by

$$P = \frac{p_1(p_1 - s)}{4(p_1 + s)} T^2. \tag{11.79}$$

If $p_1 = 2$ and $s = 1$, $t_s = \frac{1}{3}T$ and the optimal profit is $\frac{1}{3}T^2$ with shortselling and $\frac{1}{6}T^2$ without, so the prohibition of shortselling halves the maximum profit.

The solution for an arbitrary price history can be determined in this way. The procedure can best be illustrated geometrically. Suppose, for example, that $p(t)$ has the form shown in Fig. 11.3.

As well as the assumed price structure, the values of z_2 are shown in Fig. 11.3 as a broken line, both sections having the same slope s. When z_2 lies above the price curve, the trader buys, and when it is

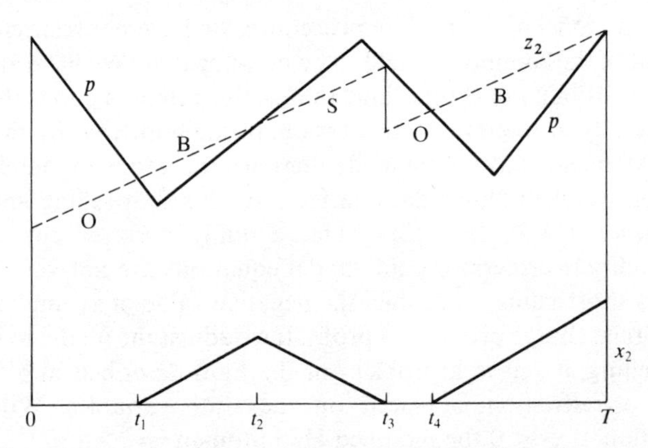

Fig. 11.3. Price variations and period of buying (B), selling (S), and no action (O).The lower line shows the amount of the commodity held by the trader.

below, there is either a period of inactivity or one of selling. There is a jump in z_2 when we pass from selling to no activity, and the value of x_2 is then zero. Hence, in this example, there are five intervals: from 0 to t_1 and from t_3 to t_4 there is no activity; buying takes place from t_1 to t_2 and from t_4 to T; and selling from t_2 to t_3. The right-hand portion of the z_2 line is fixed, but the position and magnitude of the discontinuity at t_3 have to be determined. The value of x_2 is zero between 0 and t_1 and also between t_3 and t_4, so that the position of the left-hand portion of the z_2 line is fixed by the condition that the buying and selling intervals must be of equal duration, that is, $t_2 - t_1 = t_3 - t_2$. The distribution of x_2 is shown at the bottom of Fig 11.3. By this construction we can determine all the times when there should be a change of trading activity in the optimal solution. Note that the optimal strategy for the interval 0 to t_3 is unaffected by the behaviour of the price beyond this time.

One way in which a more sophisticated model could be formed would be to allow for a feedback effect. If a large trader engages in a prolonged period of selling, the current price may be expected to fall as a consequence of his intervention. This feature could be incorporated by replacing the time-dependent price by a function of both time and selling rate. Other features that could be included would be fluctuations in exchange rates for international trade, multiple trading in more than a single commodity simultaneously, and the inclusion of accumulated interest on the money available for trading. But the simple example discussed here demonstrates how the optimal solution for commodity trading can be deduced.

11.4. Contraction of the left ventricle of the heart

As an example of a physiological system to which the ideas of optimal control can be applied, consider one phase in the action of the heart. Oxygenated blood enters the left ventricle of the heart and is expelled from it into the aorta through the aortic valve. Some of the blood leaves the aorta through branching arteries and the remainder produces a dilatation of the aorta wall. This phase of the circulation is controlled by muscular action compressing the left ventricle. The muscles are not under voluntary control, but it has been suggested that the controlling pressure has adapted to give an optimal solution for some cost function. A formalized model of the system is sketched in Fig. 11.4. During the phase of the cycle with

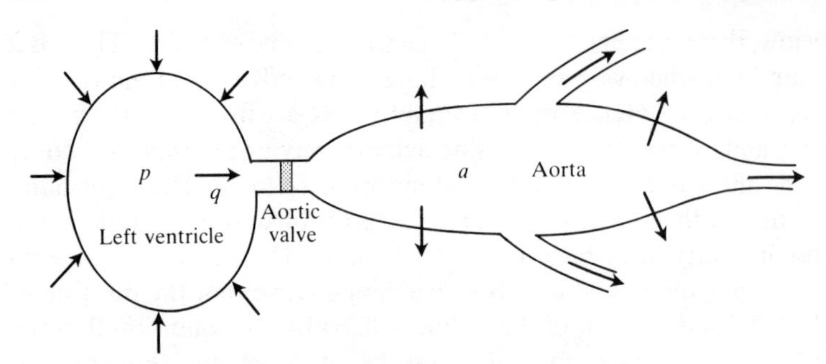

Fig. 11.4. Schematic diagram for ventricular contraction.

which we are concerned the volume $v(t)$ of the ventricle changes from V_0 at time $t = 0$ to V_1 at $t = t_1$. The rate of flow of blood from the ventricle is denoted by $q(t)$. The pressures in the aorta and in the ventricle are given by $a(t)$ and $p(t)$, respectively.

The three equations of state are as follows. The kinematic relation between the volume of the ventricle and the flow of blood gives the equation

$$\frac{dv}{dt} = -q. \tag{11.80}$$

The pressure difference between the ventricle and the aorta provides the force needed to accelerate the mass of the blood and to overcome the resistance to motion of the aortic valve. To express this balance, we have the equation

$$c_2(p - a) = \frac{dq}{dt} + c_1 q, \tag{11.81}$$

where c_1 and c_2 are positive constants. The volume of blood entering the aorta is compensated by an increase in the volume of the aorta, proportional to the rate of change of pressure there, and by the flow out of the aorta into the branching arteries at a rate proportional to the pressure. Thus we have the third state equation in the form

$$c_3 q = \frac{da}{dt} + c_4 a, \tag{11.82}$$

where c_3 and c_4 are positive constants. In these equations, we regard v, q, and a as the state variables, and p as the control variable. The initial and terminal conditions on two of the state variables are that

$$v(0) = V_0, \quad v(t_1) = V_1, \quad q(0) = q(t_1) = 0. \tag{11.83}$$

The beginning and end of this phase of the circulation are marked by zero blood flow. The condition to be applied to the aortic pressure is not so obvious. One possibility is to regard the mean value as a given constant, or, more simply, to fix the average value of the aortic pressure at the start and finish. Thus we add the condition

$$a(0) + a(t_1) = A, \tag{11.84}$$

where A is another positive constant. The terminal time t_1 is free.

To complete the specification of this problem, we must postulate a cost function. We relate the cost to the energy used by the heart muscles to produce the pressure in the ventricle and the mechanical work needed to empty the ventricle of blood. We define the cost J by the integral

$$J = \int_0^{t_1} [\tfrac{1}{2}p^2(t) + c_5 p(t)q(t)]\, \mathrm{d}t. \tag{11.85}$$

It is not easy to estimate the positive weighting constant c_5 between the two components of the cost function. The problem is now completely specified, and involves five constants, c_1 to c_5, appearing in the state and cost equations, and three parameters specifying the boundary conditions, V_0, V_1, and A. We have a linear free-time problem with a quadratic cost, although the cost function has a different form from those considered in Chapter 8, where the cost was quadratic in the state and control variables separately, and no term like the second one in (11.85) was present.

There is one novel feature in this problem, not covered by the procedure so far established. The initial state is not completely specified, since $a(0)$ is not fixed. Previously, we have nearly always assumed the initial state to be fixed, while the target could either be fixed or be made to lie in some target set. In the latter case, we derived transversality conditions which imposed some restrictions on the co-state variable; see §7.1 and §9.4. The extension of the transversality condition to cover initial states that are not completely specified was also explained in §9.4. In the present case, however, the condition (11.84) links together the initial and final values of $a(t)$. To derive the extra condition that we need to close the problem, we have to go back to the arguments presented in Chapter 9 to establish the PMP. Let $\mathbf{x}(t)$ be the optimal state vector, with $\mathbf{z}(t)$ equal to the co-state vector, and let the perturbation to the optimal solution be denoted by $\boldsymbol{\xi}(t)$. We can no longer assume that $\boldsymbol{\xi}(0) = \mathbf{0}$, since the initial state is not fixed. Similar arguments to those in §9.4 lead to the condition that

$$\mathbf{z}^{1T}\boldsymbol{\xi}^1 - \mathbf{z}^{0T}\boldsymbol{\xi}^0 = 0, \tag{11.86}$$

where the superscripts 0 and 1 denote values at $t = 0$ and at $t = t_1$, respectively. When the initial and final states are fixed, this condition reduces to an identity. When the final state lies in some target set, $\boldsymbol{\xi}^1$ must lie in that set, and (11.86) gives the transversality condition that $\mathbf{z}^1$ must be normal to the target state at the terminal point $\mathbf{x}^1$. If we apply this condition in the present case, with the state vector $\mathbf{x}$ equal to $(v, q, a)^T$, the perturbation vectors are $\boldsymbol{\xi}^{1T} = (0, 0, \xi_3^1)$ and $\boldsymbol{\xi}^{0T} = (0, 0, \xi_3^0)$, since the initial and final values of v and q are fixed by (11.83). The non-zero components of the perturbation vector are not independent since (11.84) must be satisfied, so that

$$\xi_3^0 + \xi_3^1 = 0. \tag{11.87}$$

The condition (11.86) thus becomes

$$z_3(0) + z_3(t_1) = 0, \tag{11.88}$$

where the three variables z_1, z_2, and z_3 correspond to the state variables v, q, and a, respectively.

We are now in a position to apply the PMP to this problem. Replacing the state variables v, q, and a by x_1, x_2, and x_3, respectively, and the control variable p by u_1, the Hamiltonian has the form

$$\begin{aligned} H = -\tfrac{1}{2}u_1^2 - c_5 u_1 x_2 - z_1 x_2 + z_2(-c_1 x_2 - c_2 x_3 + c_2 u_1) \\ + z_3(c_3 x_2 - c_4 x_3). \end{aligned} \tag{11.89}$$

The control variable u_1 must be chosen to maximize H, so that

$$u_1 = -c_5 x_2 + c_2 z_2, \tag{11.90}$$

and the co-state equations are

$$\dot{z}_1 = 0, \tag{11.91}$$

$$\begin{aligned} \dot{z}_2 &= c_5 u_1 + z_1 + c_1 z_2 - c_3 z_3 \\ &= -c_5^2 x_2 + z_1 + (c_1 + c_2 c_5) z_2 - c_3 z_3, \end{aligned} \tag{11.92}$$

$$\dot{z}_3 = c_2 z_2 + c_4 z_3, \tag{11.93}$$

where we have inserted the optimal value of u_1 into the second equation. The state equations in the current notation are

$$\dot{x}_1 = -x_2, \tag{11.94}$$

$$\dot{x}_2 = -(c_1 + c_2c_5)x_2 - c_2x_3 + c_2^2z_2, \tag{11.95}$$

$$\dot{x}_3 = c_3x_2 - c_4x_3. \tag{11.96}$$

Because x_1 does not appear in any of the equations, we can redefine x_1 as the change in volume, so that its initial and final values are 0 and $-V$, where $V = V_0 - V_1$. The six conditions that the solution of these six linear equations must satisfy are

$$x_1^0 = 0, \quad x_1^1 = -V, \quad x_2^0 = 0, \quad x_2^1 = 0, \quad x_3^0 + x_3^1 = A, \\ z_3^0 + z_3^1 = 0. \tag{11.97}$$

Since this is a free-time problem we must also use the condition that the Hamiltonian is zero along the optimal trajectory. Applying this condition at $t = 0$, we find that

$$\tfrac{1}{2}(c_2z_2^0)^2 - c_2z_2^0x_3^0 - c_4z_3^0x_3^0 = 0. \tag{11.98}$$

The optimal cost can be calculated in terms of the initial and final values of the state and co-state variables as in §8.2. The equivalent procedure results in the optimal cost being given by

$$J = \tfrac{1}{2}\mathbf{z}^{1\mathrm{T}}\mathbf{x}^1 - \tfrac{1}{2}\mathbf{z}^{0\mathrm{T}}\mathbf{x}^0, \tag{11.99}$$

and when the known quantities from the conditions (11.97) are inserted, we find that the optimal cost can be expressed as

$$J = \tfrac{1}{2}[-Vz_1^1 + Az_3^1]. \tag{11.100}$$

To complete the solution we have to resort to computation. The differential equations to be solved are linear, and we can integrate them forwards in time by a Runge–Kutta routine (see §12.2). Two initial conditions are prescribed by the first and third condition in (11.97), and we can construct four independent solutions by setting, in turn, one of x_3, z_1, z_2, and z_3 equal to unity and the rest to zero. The general solution is a linear combination of these solutions and, at any time τ, we can determine the solution that satisfies the four remaining conditions in (11.97). This procedure is equivalent to finding the optimal solution of the fixed-time problem with $t_1 = \tau$, so the value of H will not be zero, that is, (11.98) will not be satisfied.

We can proceed by continuing the integration of the four independent solutions to a succession of values of τ, until (11.98) is satisfied, when the optimal solution and the terminal time will be known. The optimal cost can then be calculated from (11.100).

An example of some calculations carried out using this procedure is shown in Fig. 11.5. The constants and parameters were given the following values: $c_1 = 0.3$, $c_2 = 30$, $c_3 = 0.05$, $c_4 = 0.05$, $c_5 = 2$, $A = 200$, $V = 700$. The terminal time was calculated to be $t_1 = 2.0$, and the optimal control pressure p or u_1 is shown in the figure, together with the blood flow rate q or x_2 and the aortic pressure a or x_3. The control pressure initially decreases from its starting value of 113 to a minimum of 102 at $t = 0.4$. It then rises to a maximum of 116 at $t = 1.4$ before falling rapidly to -34 at the terminal time $t_1 = 2.0$. The blood flow rate reaches a maximum value of 570 at $t = 1.6$ and the aortic pressure increases monotonically from 87 to 113, except for a small initial decrease. The variation of the controlling pressure is quite similar to that measured in a human heart, which gives some support to the view that the muscular control is designed to provide an optimal working of this part of the action of the heart.

As this example and the problem described in §11.2 have demonstrated, it is often necessary to resort to computation to arrive at the solution of an optimal control problem. Some suitable numerical

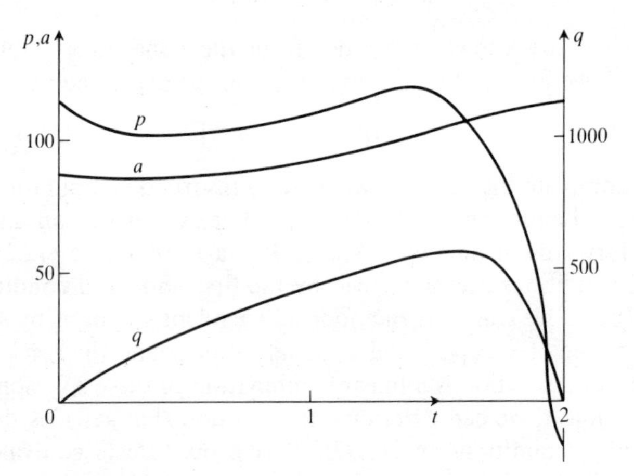

Fig. 11.5. Time variation of blood flow and the two pressures in the optimal solution.

methods are described in the final chapter, which also contains the implementation of these methods for some specific examples.

Exercises 11

(*Solutions on p. 242*)

11.1 A farmer sells a proportion u of his or her grain harvest, puts aside a fixed amount for food, and sows the remainder. The state equation and cost function have the forms

$$\dot{x} = k(x - ux - c), \quad P = \int_0^T ux \, dt,$$

and the object is to choose u to maximize P after a specified time T. The initial state is $x(0) = X$, where $X > c$. Show that the optimal strategy depends on whether T is greater or less than one and determine the optimal value of P.

In what respects does this model provide an inadequate representation of the situation?

11.2 The first controlled fishing problem of §11.1 has a state equation

$$\dot{x} = x - ux, \quad 0 \leq u \leq h.$$

Suppose that the initial stock is given by $x(0) = 1$ and that the stock is to be returned to the same level at the specified time T. Given that $h > 1$, find what choice of u will lead to the maximum catch.

11.3 Solve the second fishing problem of §11.1, with state equation

$$\dot{x} = 2x - x^2 - ux, \quad 0 \leq u \leq 2,$$

when the initial stock is very small, say $x(0) = \varepsilon$, where $0 < \varepsilon \ll 1$.

11.4 In the controlled fishing problems in §11.1 the proportion of the available stock that could be caught was given a ceiling. Suppose that instead the fishing rate is controlled. The state equation then has the form

$$\dot{x} = 2x - x^2 - u, \quad 0 \leq u \leq h.$$

Suppose that initially the stock is in the equilibrium state so that $x(0) = 2$. Find the optimal strategy to maximize the total catch in a specified time T for $h \leq 1$ and for $h > 1$.

11.5 The equation for a damped positioning problem has the form

$$\frac{d^2y}{dt^2} + \frac{1}{\varepsilon}\frac{dy}{dt} = u,$$

with $y = 1$, $dy/dt = 0$ at $t = 0$, and $y = dy/dt = 1$ at $y = T$. The terminal time T is free and the cost function is given by

$$J = \int_0^T \tfrac{1}{2}(c^2 + u^2)\,dt.$$

Find the solution for small positive values of ε by finding the solutions of the state and co-state equations in three regions: (i) an inner region near $t = 0$ in which $t = \varepsilon\tau$, (ii) a central region in which t is of order 1, (iii) a region near $t = T$ in which $t = T - \varepsilon\theta$. Show that the leading-order terms in the expansion of the optimal cost and terminal time in powers of ε are given by

$$T = \frac{1}{\varepsilon c} + 2\varepsilon + \cdots, \quad J = \frac{c}{\varepsilon} + \varepsilon c^2 \cdots.$$

(This is an example of the application of singular perturbation theory to an optimal control problem; a similar method was used in §11.2.)

11.6 Consider the trading problem of §11.3 without shortselling, with the price structure detailed in (11.73), and with $s < p_1$. Instead of starting with no stock of the commodity, suppose that $x_2(0) = X_2 > 0$. Calculate the maximum profit and compare with the profit achieved when $X_2 = 0$.

Show how the geometric construction for the solution with a general price structure can be adapted when this change in initial condition is made.

11.7 The model equations (11.64) for the trading problem do not include any interest earned on the money that is being held and not spent immediately. Show that this feature can be added by inserting an extra term into (11.64), so that is becomes

$$\dot{x}_1 = pu_1 + rx_1 - sx_2,$$

where r is the rate of interest. Discuss the changes in the solution that are introduced by this addition. In particular, calculate the maximum profit when the price is given by (11.73), with $p_1 = 2$, $s = 1$, and $r = 1$.

12 Numerical methods for optimal control problems

What would this life be without arithmetic, but a scene of horrors?

Sydney Smith

Although a great deal can be learnt from closed-form solutions of optimal control problems, in general the solution of such problems can only be obtained by computation. Some numerical methods that can be used to solve optimal control problems are discussed in this chapter.

12.1. Implementation of the Pontryagin maximum principle

Since the time-optimal principle TOP derived in Part A provides a criterion for the optimal control that is identical to that derived from the PMP as applied to time-optimal problems (see §9.1), we need only discuss the way in which the PMP can be implemented in a numerical solution of a general optimal control problem. The development of appropriate algorithms for such problems is fraught with obstacles, and it does not seem possible to produce schemes that will apply in general. Each problem, or class of problems, has to be considered on its own; all that can be done here is to describe methods that work in some specific cases.

There are various difficulties that have to be overcome in attempting a solution of an optimal control problem in any other than the simplest cases. The state and co-state equations cannot be solved as an initial-value problem, and, because of the nonlinearity, an iterative method must be used to ensure that the terminal conditions are satisfied. The choice of the optimal control depends on the values of the state and co-state variables, and is to be found by maximizing the Hamiltonian. This also is a nonlinear problem whose solution requires some form of iteration on the m control variables. For this iteration to succeed, the starting point of the iteration may have to be very close to the true solution, and there is the added complication that the iteration may converge to an alternative solution. We know that some quite simple problems have more than one extremal solution, and it is only when all these have been found that we can identify the optimal one. This difficulty is accentuated for nonlinear problems since we do not know how many possible candidates for the optimal solution there may be. Yet another difficulty arises when the terminal time is unknown. Discretization of the state, co-state, and control variables defines them by their values at equally spaced time intervals; the differential equations are then replaced by non-linear algebraic equations. This is a useful method for a given time interval, but is not so convenient when the final time is unknown.

Because of these and other difficulties, there does not seem to be any hope of developing a general method for finding a solution of an arbitrary optimal control problem. Instead, some particular cases will be discussed in detail in order to show by example how some of these difficulties can be overcome. Before doing so, however, it is convenient to describe briefly some of the standard numerical techniques that will be used.

12.2. Numerical techniques

Consider a system of n first-order differential equations of the form

$$\dot{\mathbf{x}} = \mathbf{f}(\mathbf{x}, t), \tag{12.1}$$

with $\mathbf{x}(t_0) = \mathbf{x}^0$. Their forward integration in time can be found by applying a *Runge–Kutta method*; these methods are based on the Taylor expansion of the solution and are included in most packages of computer algorithms. They are particularly useful when the functions $\mathbf{f}$ are easy to evaluate. The application of a Runge–Kutta method over a single step of length h replaces the variables at time t by their values at time $t + h$, and successive applications produce a solution for the initial-value problem.

For a boundary-value problem, Runge–Kutta methods can still be used, but they must be combined with an iterative search. Suppose that the initial and final conditions are given by

$$x_i(t_0) = x_i^0 \quad \text{for } m + 1 \le i \le n, \quad x_i(t_1) = x_i^1 \quad \text{for } 1 \le i \le m, \tag{12.2}$$

so that there are m unknown components of $\mathbf{x}$ at the initial time. We have to choose these m unknowns in such a way that the m

conditions at $t = t_1$ are satisfied. For linear equations, we can find m independent solutions and so obtain m linear equations for the unknowns, which can then be found by *Gaussian elimination.* If the equations are written in matrix form as $\mathbf{Ax} = \mathbf{b}$, row operations on the matrix can be used to replace all the elements below the leading diagonal by zero. Then, starting from the last row and working upwards, we can find the values of the components of $\mathbf{x}$ successively. For nonlinear equations there is no foolproof and systematic method. The best we can do is to take two sets of initial values. Solving the equations gives two sets of final values and we can use linear interpolation to find the appropriate combination of initial values that would give the correct final value if the mapping were a linear one. This procedure can be repeated with one of the original sets of values replaced by the interpolated set. If the initial guesses are not too far from the true solution, there is some possibility that continuing in this manner will produce a convergent sequence. This technique employs the m-dimensional version of the *secant method* for solving nonlinear algebraic equations.

An alternative search procedure is to use the *bisection method.* Suppose we do not know the initial value of z_1 and that the target is $x_1 = x_2 = 0$. We begin by trying a set of initial values for $z_1(0)$ until we find two, say A_1 and A_2, for which the corresponding values of x_1 when $x_2(t) = 0$ are B_1 and B_2, where B_1 and B_2 have opposite signs. We expect the correct value of $z_1(0)$ to lie between A_1 and A_2, so we try $A_3 = \frac{1}{2}(A_1 + A_2)$ and solve the equations to find the corresponding value of B_3. We repeat the process with $A_4 = \frac{1}{2}(A_1 + A_3)$ if B_1 and B_3 have opposite signs and with $A_4 = \frac{1}{2}(A_2 + A_3)$ if B_2 and B_3 have opposite signs. The same criterion as used with the secant method determines when the process can be terminated.

In applying the PMP to determine the optimal control, we need to be able to find the supremum of a function H of the m control variables $\mathbf{u}$. A local maximum can be found by solving the equations

$$g_i \equiv \frac{\partial H}{\partial u_i} = 0, \quad i = 1, 2, \ldots, m. \tag{12.3}$$

To solve these nonlinear equations, we can employ *Newton iteration,* since it is usually possible to evaluate the derivatives of g_i. If $\mathbf{u}$ is an approximate solution of (12.3), then $\mathbf{u} + \mathbf{v}$ is a better approximation, where $\mathbf{v}$ satisfies the equations

$$\mathbf{Gv} = -\mathbf{g}. \tag{12.4}$$

G is the matrix with elements equal to $\partial g_i(\mathbf{u})/\partial u_j$ and **g** is the vector with components $g_i(\mathbf{u})$. A useful feature of Newton iteration is that in general it converges rapidly, but it does require that the initial approximate solution be sufficiently close to the required solution. Also, there may be several values of **u** that give stationary values of H and the search for the value that gives the supremum of H may be difficult to accomplish. A further difficulty arises if there are bounds to the values of some or all of the control variables. Then it is not enough to determine a local maximum of H by finding its stationary points; the values of H with **u** on the boundary of the control set must also be examined.

An alternative procedure to the Runge–Kutta methods for solving differential equations that does not suffer from the restriction to initial-value problems is the technique of *discretization*. Consider first the single differential equation

$$\dot{x} = f(x, t) \quad \text{for } t_0 \leq t \leq t_1. \tag{12.5}$$

We divide the time interval into N equal parts of duration $\delta t = (t_1 - t_0)/N$ and label the successive instants by τ_j, $j = 0, 1, \dots, N$, where $\tau_j = t_0 + j\delta t$. The numerical solution of the differential equation will have been achieved when we have found the values of x at these instants. If we write $x(\tau_j) = x_j$, we can replace (2.5) by the finite-difference approximations

$$x_{j+1} - x_j = \tfrac{1}{2}\delta t[f(x_j, \tau_j) + f(x_{j+1}, \tau_{j+1})]. \tag{12.6}$$

This approximation is obtained by evaluating the terms in (12.5) at the time $\tau_j + \frac{1}{2}\delta t$ and is accurate to the second order in δt. These equations, for $j = 0, 1, \dots, N - 1$, together with one boundary condition are a set of $N + 1$ algebraic equations for the values of x_j. In general these equations will be nonlinear and to solve them we must use Newton iteration, with Gaussian elimination to solve the set of linear equations obtained at each step in the iterative procedure.

The extension of this method to a system of n equations is straightforward. Note that in this method the unknowns are determined simultaneously and not successively as in the Runge–Kutta methods, and that initial- and boundary-value problems can be solved equally well.

Further details and justification for these techniques can be found in many texts on numerical methods, some of which are listed in the Bibliography. We can now proceed to consider the application of these methods to the solution of some specific optimal control problems.

12.3. Optimal controls determined explicitly: discretization

The numerical solution of an optimal control problem is greatly simplified when the PMP determines the optimal control as an explicit function of the state and co-state variables. Such is the case, for example, in the linear–quadratic problems with unbounded controls studied in Chapter 8. The Hamiltonian is a quadratic function of **u** and its local maximum suffices to determine **u** explicitly and uniquely. The problem is thus reduced to that of solving a set of $2n$ linear differential equations. Because the initial values of the co-state variables are unknown, it is not an initial-value problem and discretization is more appropriate than a Runge–Kutta method. The only remaining difficulty occurs when the terminal time is not fixed; to compensate for this loss of information we have to use the fact that the Hamiltonian is zero along the optimal trajectory.

This class of problems can be extended to include nonlinear state equations, provided they remain linear in the control variables. As an illustration of how appropriate techniques for these problems can be constructed, let us consider a one-dimensional problem, with state equation and cost function given by

$$\dot{x} = 1 - x + xu, \quad J = \int_0^{t_1} \tfrac{1}{2}(u^2 + x^2)\,dt. \tag{12.7}$$

Suppose first that the terminal time t_1 is fixed, and that the initial and final conditions are given

$$x(0) = 0, \quad x(t_1) = 1. \tag{12.8}$$

The Hamiltonian is given by

$$H = -\tfrac{1}{2}u^2 - \tfrac{1}{2}x^2 + z(1 - x + xu), \tag{12.9}$$

so that the optimal control u is determined explicitly and uniquely by finding the supremum of H. With $u = xz$ the state and co-state equations have the forms

$$\dot{x} = 1 - x + x^2 z, \quad \dot{z} = x + z - xz^2. \tag{12.10}$$

If $\hat{x}$ and $\hat{z}$ are approximate values for these variables, we can obtain the Newton improvements by writing $x = \hat{x} + (x - \hat{x})$ and expressing z similarly. After substituting these forms into (12.10) and neglecting second-order terms we obtain the linear equations

$$\begin{aligned} \dot{x} &= 1 - x + 2\hat{x}\hat{z}x + \hat{x}^2 z - 2\hat{x}^2\hat{z}, \\ \dot{z} &= x - \hat{z}^2 x + z - 2\hat{x}\hat{z}z + 2\hat{x}\hat{z}^2. \end{aligned} \tag{12.11}$$

If we use a subscript j to denote the value of a variable at the time jh, where $h = t_1/N$, the finite-difference forms of these equations can be written as

$$\begin{aligned} a_j x_j + b_j z_j + c_j x_{j+1} + d_j z_{j+1} &= r_j, \\ A_j x_j + B_j z_j + C_j x_{j+1} + D_j z_{j+1} &= R_j, \end{aligned} \tag{12.12}$$

for $j = 0, 1, \ldots, N-1$, where

$$\begin{aligned} &a_j = -1 + \tfrac{1}{2}h - h\hat{x}_j\hat{z}_j, \quad b_j = -\tfrac{1}{2}h\hat{x}_j^2, \\ &c_j = 1 + \tfrac{1}{2}h - h\hat{x}_{j+1}\hat{z}_{j+1}, \quad d_j = -\tfrac{1}{2}h\hat{x}_{j+1}^2, \\ &A_j = -\tfrac{1}{2}h + \tfrac{1}{2}h\hat{z}_j^2, \quad B_j = -1 - \tfrac{1}{2}h + h\hat{x}_j\hat{z}_j, \\ &C_j = -\tfrac{1}{2}h + \tfrac{1}{2}h\hat{z}_{j+1}^2, \quad D_j = 1 - \tfrac{1}{2}h + h\hat{x}_{j+1}\hat{z}_{j+1}, \\ &r_j = h(1 - \hat{x}_j^2\hat{z}_j - \hat{x}_{j+1}^2\hat{z}_{j+1}), \quad R_j = h(\hat{x}_j\hat{z}_j^2 + \hat{x}_{j+1}\hat{z}_{j+1}^2). \end{aligned} \tag{12.13}$$

In addition, we have the initial and final conditions, which are $x_0 = 0$ and $x_N = 1$. If we write the unknowns as a vector $\mathbf{X}$ of dimension $2N + 2$, with components $x_0, z_0, x_1, z_1, \ldots$, these equations can be written in matrix form as $\mathbf{AX} = \mathbf{S}$, where $\mathbf{A}$ is a five-banded matrix, with non-zero elements occupying the leading diagonal and the two diagonals above and below it. The matrix $\mathbf{A}$ is in a more suitable form for computation if each pair of equations (12.12) is written down in the opposite order; the vector $\mathbf{S}$ then contains the right-hand sides of the equations in the order $0, R_0, r_0, \ldots, R_{N-1}, r_{N-1}, 1$.

In order to initiate this iterative process, it is necessary to begin with a choice of the values of $\hat{x}_j$ and $\hat{z}_j$. Unfortunately there is no automatic way to determine these starting values. If an initial guess does not lead to a converging sequence, the only thing to do is to start again with a different choice. After each iteration has been completed, it is possible to evaluate the cost by using the trapezium rule to evaluate the integral. The sequence of values of J so obtained can be used to judge the convergence of the iteration, which is terminated when successive values differ by less than some previously fixed tolerance. The whole calculation should then be repeated for a larger value of N until one is satisfied that a sufficiently accurate result has been obtained. It is advantageous to begin with N not large. Then the search for a suitable initial guess may be expedited, and when a converged solution has been found, it can be used as the basis for the starting values for a larger value of N.

The procedure described above has been written as a computer program. With $t_1 = 1$ and $N = 10$ and with the initial values $x = t$, $z = 0$, the values of the cost J as a function of the number of iterations is shown in Table 12.1. The tolerance used was 10^{-5}. The values for $N = 20$ and $N = 40$ are also shown. The initial choices for x and z for the higher values of N were interpolated from the converged values already determined. The cost at the iteration number 0 denotes the cost as calculated from the chosen initial values.

The error is expected to be of order N^{-2} and this is confirmed by the results; if we write $J(N)$ for the cost as calculated using N intervals, $J(40) - J(20)$ is approximately one-quarter of the value of $J(20) - J(10)$. We can therefore extrapolate from the calculated results to find an estimate J_{est} of the correct value of J in the form

$$J_{\text{est}} = \frac{4J(40) - J(20)}{3} = 1.018\,403.$$

This example is one when the terminal time is fixed. Suppose we consider the same problem, but with the value of the terminal time t_1 unknown. We can try to find a solution by discretization, postulating the number of intermediate values of t; that is, we can choose N but

Table 12.1. Optimal cost for the problem defined by (12.7) and (12.8) with fixed time.

Iteration number	$N = 10$	20	40
0	0.335 000	1.016 440	1.017 902
1	0.809 740	1.017 867	1.018 276
2	1.027 190	1.017 902	1.018 278
3	1.016 370	1.017 902	
4	1.016 440		
5	1.016 440		

do not know the value of h, which has to be found. Instead we know that the Hamiltonian must be zero, and this condition, applied at $t = 0$, shows that $z(0) = 0$. The matrix $\mathbf{A}$ and the vector $\mathbf{S}$ are linear functions of h, so we can write the equation $\mathbf{AX} = \mathbf{S}$ in the form

$$(\mathbf{A}^1 + h\mathbf{A}^2)\mathbf{X} = \mathbf{S}^1 + h\mathbf{S}^2. \tag{12.14}$$

Note that the superscripts do not denote powers. If $\hat{h}$ and $\hat{\mathbf{X}}$ are approximate solutions of these equations, Newton iteration results in linear equations for the improved solutions in the form

$$(\mathbf{A}^1 + \hat{h}\mathbf{A}^2)\mathbf{X} = \mathbf{S}^1 + h(\mathbf{S}^2 - \mathbf{A}^2\hat{\mathbf{X}}) + \hat{h}\mathbf{A}^2\hat{\mathbf{X}}. \tag{12.15}$$

Although h is still to be determined, we can solve these equations as a linear function of h in the general form

$$\mathbf{X} = \mathbf{X}^1 + h\mathbf{X}^2. \tag{12.16}$$

The value of $z(0)$ forms the second component of the vector $\mathbf{X}$, so the condition $z(0) = 0$ which ensures the vanishing of the Hamiltonian is satisfied provided

$$h = -X_2^1/X_2^2. \tag{12.17}$$

Of course, the two iteration schemes required by the nonlinear differential equations and the unknown terminal time should be carried out concurrently. The iteration is terminated when both the cost and the terminal time have been calculated to within assigned tolerances. Numerical results for this problem are given in Table 12.2, where the values of J are given for various values of N and the tolerance is 10^{-5}.

Since the error is again of order N^{-2}, we can extrapolate in the same way as before. The extrapolated value of the optimal cost J_{est} is

Table 12.2. Optimal cost for the problem defined by (12.7) and (12.8) with t_1 not fixed.

Iteration number	$N = 10$	20	40
0	0.335 000	0.708 014	0.704 520
1	0.942 486	0.704 520	0.703 636
3	0.708 031	0.704 522	0.703 636
4	0.708 014		
5	0.708 014		

given by $J_{\text{est}} = 0.703\,341$ and the terminal time extrapolated in the same way is given by $t_{1\,\text{est}} = 1.451\,600$. Note that the optimal cost when the terminal time is not prescribed is about 30 per cent lower than that incurred when the terminal time was fixed at $t_1 = 1$.

12.4. Optimal controls determined explicitly: Runge–Kutta method

As an application of a Runge–Kutta method, with an unknown terminal time, consider the following nonlinear variant of the positioning problem. We suppose that the state equations and cost function are given by

$$\dot{x}_1 = x_2(1 + x_1^2), \quad \dot{x}_2 = u, \quad J = \int_0^{t_1} (\tfrac{1}{2}u^2 + 2)\,\mathrm{d}t, \tag{12.18}$$

and that the initial and final states are given by

$$x_1(0) = 1, \quad x_2(0) = 0, \quad x_1(t_1) = 0, \quad x_2(t_1) = 0. \tag{12.19}$$

The Hamiltonian for this problem is given by

$$H = -2 - \tfrac{1}{2}u^2 + z_1 x_2(1 + x_1^2) + z_2 u, \tag{12.20}$$

so the PMP show that the optimal control is given by $u = z_2$, and, since the Hamiltonian must be zero at $t = 0$, $z_2(0) = \pm 2$. Because we want x_1 to decrease, we expect x_2 to be negative initially, which suggests that we should choose $z_2(0) = -2$. The state and co-state equations are then given by

$$\dot{x}_1 = x_2(1 + x_1^2), \quad \dot{x}_2 = z_2, \quad \dot{z}_1 = -2x_1x_2z_1, \quad \dot{z}_2 = -z_1(1 + x_1^2), \tag{12.21}$$

and the initial values of x_1, x_2, and z_2 are known. The solutions of these equations form a one-dimensional family, depending on the initial value of z_1, and we are seeking the solution that passes through the origin at some unknown time t_1.

Suppose that A_1 and A_2 are two values for $z_1(0)$ and that we solve (12.21) for increasing t until $x_2 = 0$. If B_1 and B_2 are the corresponding values of x_1 at this instant, the secant method predicts that a better solution will be found by using the initial value $z_1(0) = A_3$, where

$$A_3 = \frac{A_1B_2 - A_2B_1}{B_2 - B_1}. \tag{12.22}$$

Table 12.3. Terminal time and optimal cost for the problem defined by (12.8) and (12.19), using a Runge–Kutta method.

A	t_1	J
−1.000 00	2.000 00	8.000 00
−1.500 00	1.333 23	8.665 99
−1.378 75	1.450 58	8.416 11
−1.286 91	1.554 06	8.225 56
−1.304 25	1.533 35	8.283 41
−1.302 85	1.534 99	8.281 07
−1.302 82	1.535 04	8.281 01

We solve the equations again, with this initial value, and find that $x_1 = B_3$, say, when $x_2 = 0$. We can then reapply the secant method to find A_4 from the pairs A_3, B_3 and A_2, B_2. This procedure continues until a value of $z_1(0)$ is found for which x_1 differs from zero by less than an assigned tolerance at the instant when $x_2 = 0$.

Some results obtained by this method for the problem defined by (12.18) and (12.19) are given in Table 12.3. The successive values of A are shown in the first column, and the corresponding values of the time t_1 and the cost J are in the second and third columns. The tolerance was taken to be 10^{-4} and the step length was $h = \frac{1}{40}$.

The optimal values estimated by extrapolation from these results and those with $h = \frac{1}{20}$ are $t_{1\ \mathrm{est}} = 1.535\ 00$ and $J_{\mathrm{est}} = 8.281\ 61$.

12.5. Optimal controls determined conditionally

The PMP does not always determine the optimal control unconditionally. If the controls are bounded the supremum of H may occur at a local maximum or when the control vector lies on the boundary of the control set. Which of these two possibilities occurs will depend on the values attained by the state and co-state variables. In a number of examples that have been discussed in the preceding chapters, the optimal control has been discontinuous and to find the solution we have had to consider a variety of possibilities before the optimal solution could be identified. The corresponding numerical procedure is not easy to formulate in general terms. Because of the conditional choice of the control, discretization methods are not well suited to these problems.

We look first at a problem in which the optimal control is continuous but takes its extreme values for part of the time. Suppose we consider again the problem discussed in §12.4, but with the added restriction that $|u| \leq 1$. The supremum of H occurs when $z_2u - \frac{1}{2}u^2$ has its maximum value and this means that $u = z_2$, provided $|z_2| \leq 1$. If $z_2 > 1$, we must choose $u = 1$ and if $z_2 < -1$, $u = -1$. From the value of H at $t = 0$, we deduce that we should begin the calculation with $u = -1$ and $z_2(0) = -2.5$. The equations to be solved are the same as those in (12.21), except for the second one, which now has the form

$$\dot{x}_2 = z_2, \quad \text{when } |z_2| \leq 1, \quad \dot{x}_2 = \text{sgn}(z_2), \quad \text{when } |z_2| > 1. \quad (12.23)$$

The calculation can then proceed by using a Runge–Kutta method in the same way as explained in §12.4. Some results are presented in Table 12.4, using the same tolerance and step length; note that the optimal cost is greater than when there is no restriction on the size of the control, as expected. The terminal time, however, is reduced. The optimal control took its extreme values for the intervals $0 < t < 0.411\,90$ and for $0.961\,09 < t < 1.372\,95$ and varied smoothly between -1 and $+1$ for $0.411\,90 < t < 0.961\,09$.

Results extrapolated from these obtained using $h = \frac{1}{20}$ and $h = \frac{1}{40}$ are found to be given by $t_{1\ \text{est}} = 1.372\,88$ and $J_{\text{est}} = 11.8521$.

In some of the problems with bounded controls discussed in Chapters 6 and 7 the optimal control was found to be discontinuous. In time-optimal problems the optimal control is bang–bang, with each component only taking its extreme values. Let us now consider

Table 12.4. Optimal time and cost for the same problem as in Table 12.3, but with bounded controls.

A	t_1	J
−1.300 00	1.923 05	10.3460
−1.500 00	1.666 58	10.8328
−1.641 32	1.523 13	11.2527
−1.766 58	1.415 06	11.6623
−1.812 19	1.379 48	11.8195
−1.820 40	1.373 29	11.8484
−1.820 86	1.327 94	11.8500
−1.820 85	1.372 95	11.8499

a nonlinear problem for which we expect the optimal control to vary discontinuously.

The state equations for a certain nonlinear oscillator with a single control can be written in the form

$$\dot{x}_1 = x_2, \quad \dot{x}_2 = -x_1(1 + cx_1^2) + u, \tag{12.24}$$

where c is a constant. The linear problem with $c = 0$ was considered in §5.3. We suppose that we wish to optimize the time taken to reach the target and that the control is bounded by U, so that

$$J = \int_0^{t_1} 1 \, \mathrm{d}t, \quad -U \le u \le U. \tag{12.25}$$

The terminal time is free, and the initial and final states are given by

$$x_1(0) = X_1, \quad x_2(0) = X_2, \quad x_1(t_1) = 0, \quad x_2(t_1) = 0. \tag{12.26}$$

The supremum of the Hamiltonian, defined by

$$H = -1 + z_1 x_2 + z_2 u - z_2 x_1(1 + cx_1^2), \tag{12.27}$$

is attained for values of the control u defined by

$$u = U \quad \text{when } z_2 > 0, \quad u = -U \quad \text{when } z_2 < 0 \tag{12.28}$$

and the co-state equations are given by

$$\dot{z}_1 = z_2(1 + 3cx_1^2), \quad \dot{z}_2 = -z_1. \tag{12.29}$$

Suppose we concentrate on the special cases in which $c > 0$, $X_1 > 0$, and $X_2 = 0$. Then, if we apply the condition that $H = 0$ at $t = 0$, we find that it is possible to begin with $u = -U$ for all values of X_1, provided

$$z_2(0) = \frac{-1}{U + X_1(1 + cX_1^2)}. \tag{12.30}$$

It is also possible to begin with $u = U$ for values of X_1 such that $X_1(1 + cX_1^2) < U$ with

$$z_2(0) = \frac{1}{U - X_1(1 + cX_1^2)}. \tag{12.31}$$

The most convenient method for obtaining the solution, however, is to commence the calculation at the target and work backwards in time. We do not know in advance whether to begin with u positive or negative, but it is easy to see that reversing the sign of the optimal

Table 12.5. Optimal time and number of switches required for a nonlinear oscillator.

c	Optimal time	Number of switches
0.0	15.71	5
1.0	18.45	8
2.0	21.22	10
3.0	23.36	12
4.0	24.89	13
5.0	27.03	15

control in any trajectory that leads to $x_1 = X_1$ will lead to $x_1 = -X_1$, so that we can start with $u = U$ and end the calculation when $|x_1| = X_1$. The value of H at $t = t_1$ shows that $z_2(t_1) = 1$, given that $u(t_1) = U$, but $z_1(t_1)$ is unknown. For each chosen value of this unknown, we can proceed with the calculation to determine a succession of values of x_1 when x_2 passes through zero, ending each run when $|x_1|$ exceeds X_1. We then repeat the procedure for a different initial value for z_1 and apply the bisection method until we find a trajectory that passes through $x_1 = X_1$, $x_2 = 0$.

The optimal times for various values of c are shown in Table 12.5, for $X_1 = 1.0$ and $U = 0.1$. The number of switches made by the optimal control between its extreme values is also shown.

The numerical methods described and used in this chapter are adequate to determine the solutions of optimal control problems with bounded or unbounded controls and with linear or nonlinear state equations. There is no way of guaranteeing that they will be equally successful in general, especially when there are several control variables. Each problem must be examined afresh, but the methods described and the way in which they can be implemented should provide some guidance in the development of an appropriate algorithm for a wide variety of problems. As is often the case in applied numerical analysis, experience leads to expertise.

12.6. Conclusion

Further examples of the application of the theory developed in this book can be found in some of the texts referenced in the Bibliography. There are many extensions of the theory of optimal control beyond the introductory material that has been presented here, some of which are covered by books listed in the Bibliography. Controls may be applied at discrete times and not continuously. Multi-dimensional systems may be governed by partial differential equations, not ordinary ones. The deterministic systems which we have studied may be replaced by stochastic ones. The purpose of this book has been to discuss extensively one special form of the theory and to show how relatively simple mathematics can be used to solve a variety of optimal control problems covering many different fields of interest.

Exercises 12

(*Solutions on p. 247*)

12.1 Write a computer program to solve the equations developed in §11.4 for the heart problem. Examine how the optimal solution is affected when one or other of the five coefficients c_1 to c_5 is put equal to zero. (For example, if $c_1 = 0$ there is no pressure drop across the aortic valve.) In this manner, determine which, if any, of the features included in the model can be omitted to give a simpler description that still adequately represents the physical situation.

12.2 Compute the solutions for some of the problems solved analytically in the chapters of this book. The known solutions will provide a check on the calculations.

Bibliography

Some books are to be tasted, others to be swallowed, and some few to be chewed and digested.

Francis Bacon

Listed here is a selection of books on optimal control theory that are relevant to the material covered in this book. It is followed by references with particular connections to individual chapters and by details of some books concerned with applications to specific fields.

The pioneering book on optimal control theory is

Pontryagin, L. S., Boltyanskii, V. G., Gamkrelidze, R. S., and Mishchenko, E. F. (1962). *The mathematical theory of optimal processes.* Interscience, New York.

Other books with comprehensive coverage are

Athans, M. and Falb, P. L. (1966). *Optimal control: An introduction to the theory and its applications.* McGraw-Hill, New York; Leitmann, G. (1966). *An introduction to optimal control.* McGraw-Hill, New York; Lee, E. B. and Markus, L. (1967). *Foundations of Foundations of optimal control theory.* Wiley, New York; Bell, D. J. and Jacobson, D. H. (1975). *Singular optimal control problems.* Academic Press, London.

A book that includes more recent material is

Alekseev, V. M., Tikhomirov, V. M., and Fomin, S. V. (1987). *Optimal control.* Plenum, New York.

A mathematically more advanced text is

Berkovitz, L. D. (1974). *Optimal control theory.* Springer-Verlag, New York.

Two introductory texts, which also contains some applications, are

Knowles, G. (1981). *An introduction to applied optimal control.* Academic Press, New York; Macki, J. and Strauss, A. (1982). *Introduction to optimal control theory.* Springer-Verlag, New York.

The book by Macki and Strauss is a useful text for consultation by those who wish to see the mathematical arguments presented in an elementary fashion but with precision.

The extension to stochastic systems is described in an elementary way in

Bryson, A. E., Jr and Ho, Y.-C. (1967). *Applied optimal control.* Blaisdell, Waltham, MA,

and in a comprehensive fashion in

Grimble, M. J. and Johnson, M. A. (1988). *Optimal control and stochastic estimation*, 2 Vols. Wiley, Chichester.

The following books have particular relevance to the specified chapters of this book.

Chapter 1. Techniques and examples of mathematical modelling in deterministic systems are described in

Lin, C. C. and Segal, L. A. (1974). *Mathematics applied to deterministic problems in the natural sciences.* Macmillan, New York; Tayler, A. B. (1986). *Mathematical models in applied mechanics.* Oxford University Press.

Chapter 2. More details concerning the solution of differential equations and linear algebra of matrices can be found in most mathematics textbooks written for undergraduate scientists and engineers. See, for example,

Stephenson, G. (1973). *Mathematical methods for science students.* Longman, London; Grossman, S. I. and Derrick, W. R. (1988). *Advanced engineering mathematics.* Harper & Row, New York.

A more advanced treatment for equations with many variables can be found in

Thurston, H. (1989). *Intermediate mathematical analysis.* Oxford University Press.

For the properties of convex sets, see

Eggleston, H. G. (1969). *Convexity.* Cambridge University Press; Valentine, F. A. (1964). *Convex sets.* McGraw-Hill, New York.

Chapter 3. Control theory is the natural preliminary to optimal control theory. The classic text on the mathematical theory is

Bellman, R. E. *Introduction to the mathematical theory of control processes.* (Vol. 1, 1967; Vol. 2, 1971) Academic Press, New York.

A recent introduction to the subject is

Barnett, S. and Cameron, R. G. (1985). *Introduction to mathematical control theory*, (2nd edn). Oxford University Press.

Chapter 6. The apparatus of the theory of the calculus of variations has not been used here, although it is closely connected with the idea of optimization. A brief account of the theory can be found in

Arthurs, A. M. (1975). *Calculus of variations.* Routledge & Kegan Paul, London.

A more extensive discussion is provided by

Pars, L. A. (1962). *Calculus of variations.* Heinemann, London.

Chapter 11. Further information about the fishery problems of §11.1 and related questions can be found in

Clark, C. W. (1976). *Mathematical bioeconomics.* Wiley, New York.

For the theory of epidemics, see

Waltman, P. (1974). *Deterministic threshold models in the theory of epidemics.* Springer-Verlag, New York.

For information about the techniques of singular perturbations, as used in §11.2, see

Nayfeh, A. H. (1981). *Introduction to perturbation techniques.* Wiley, New York.

For the application of perturbation methods to optimal control problems, see

Bensoussan, A. and Inria (1988). *Perturbation methods in optimal control.* Wiley, Chichester.

Background material for applications to economics can be found in

Allen, R. G. D. (1966). *Mathematical economics.* Macmillan, New York; Intriligator, M. D. (1971). *Mathematical optimization and economic theory.* Prentice-Hall, Englewood Cliffs, NJ.

Management problems, of which §11.3 is an example, are discussed in

Benoussan, A., Hurst, E., and Naslund, B. (1974). *Management applications of modern control theory.* North-Holland, New York.

The model for the heart problem in §11.4 is derived from

Noldus, E. J. (1976). Optimal control aspects of left ventricular ejection dynamics. *Journal of Theoretical Biology*, **63**, 275.

Chapter 12. The numerical methods described briefly in this chapter, as well as many others, are treated extensively in most standard texts on numerical analysis. See, for example,

Ralston, A. and Rabinowitz, P. (1978). *A first course in numerical analysis,* (2nd edn). McGraw-Hill, London; Press, W. H., Flannery, S. A., Teukolsky, S. A., and Vetterling, W. T. (1986). *Numerical recipes: The art of scientific computing.* Cambridge University Press.

A general numerical method for optimal control problems, based on their restatement as problems in the calculus of variations, is given by

Gregory, J. and Lin, C. (1989). Efficient general methods in optimal control theory. In *Modern optimal control* (ed. E. O. Roxin). Marcel Dekker, New York.

The following are some books that contain applications of optimal control theory to a variety of systems. The medical problems described in Chapters 1, 6, 8, and 11, as well as many others, are to be found in

Swan, G. W. (1984). *Applications of optimal control theory in biomedicine.* Marcel Dekker, New York.

Applications to electric power supply are in

Christensen, G. S., El-Hawary, M. E., and Soliman, S. A. (1987). *Optimal control applications in electric power systems.* Plenum, New York;

applications to plasma physics are in

Blum, J. (1989). *Numerical simulation and optimal control in plasma physics, with applications to tokamaks.* Wiley, Chichester;

and the problems of oil recovery are described in

Ramirez, W. F. (1987). *Application of optimal control theory to enhanced oil recovery.* Elsevier, Amsterdam.

Solutions to the exercises

I told my secret out, That none might be in doubt.
Robert Bridges

Exercises 1 (p. 12)

1.1 With $u = 3t + v$, the solution of the state equation is $x = t^3 + \int_0^t vt_1\, dt_1$, so the condition $x(1) = 1$ gives the required result.
With $u = 3t + v$, the cost $J = 3 + \int_0^1 6vt\, dt + \int_0^1 v^2\, dt$ and hence $J \geq 3 = J^*$, with equality only when $v(t) = 0$. Thus u^* is the optimal control.

1.2 Starting with $u_1 = -1$, we have $x_2 = X - t$, $x_1 = Xt - \frac{1}{2}t^2$. With t_1 the terminal time, the solution after switching has $u_1 = 1$ and goes through the origin, so that $x_2 = t - t_1$, $x_1 = \frac{1}{2}(t - t_1)^2$. Equating the two values of x_1 and those of x_2 at the switching time t_2, we arrive at two equations with solution $t_2 = X(1 + \frac{1}{2}\sqrt{2})$, $t_1 = X(1 + \sqrt{2})$. For this to be an acceptable solution we must have $0 \leq t_2 \leq t_1$, which is clearly satisfied when $X > 0$.

1.3 See, for example, the problems discussed in Chapter 11.

Exercises 2 (p. 27)

2.1 From the given form of $x(t)$ in (2.11), it follows that $x(t_0) = X(t_0, t_0)x^0 = x^0$. Also, differentiating (2.11) gives

$$\dot{x} = A(t)X(t, t_0)\left(x^0 + \int_{t_0}^{t} X(t_0, \tau)b(\tau)\, d\tau\right) + X(t, t_0)X(t_0, t)b(t),$$

which verifies that (2.8) is satisfied since $X(t, t_0)X(t_0, t) = 1$.

2.2 The solutions of the state equations with $x_1(t_0) = 1$, $x_2(t_0) = 0$ are $x_1 = \exp(t - t_0)$, $x_2 = 0$, and the solutions with $x_1(t_0) = 0$, $x_2(t_0) = 1$ are $x_2 = \exp(t - t_0)$, $x_1 = \frac{1}{2}(t^2 - t_0^2)\exp(t - t_0)$. Hence

$$\mathbf{X}(t, t_0) = \begin{bmatrix} \exp(t - t_0) & \frac{1}{2}(t^2 - t_0^2)\exp(t - t_0) \\ 0 & \exp(t - t_0) \end{bmatrix}.$$

The solution with the given values at $t = 0$ is given by $\mathbf{x} = \mathbf{X}(t, 0)\mathbf{c}$, where $\mathbf{c}$ is the column with elements c_1 and c_2. Hence $x_1 = c_1 e^t + \frac{1}{2}c_2 t^2 e^t$, $x_2 = c_2 e^t$.

2.3 The fundamental matrices $\mathbf{X}(t, 0) = \exp(\mathbf{A}t)$ are most easily found by calculating $\mathbf{A}^n$ in each case. The answers are

$$\text{(a)}\quad \mathbf{X} = \begin{bmatrix} e^t & e^t - 1 \\ 0 & 1 \end{bmatrix}, \quad \text{(b)}\quad \mathbf{X} = \begin{bmatrix} \exp(pt) & t\exp(pt) \\ 0 & \exp(pt) \end{bmatrix},$$

$$\text{(c)}\quad \mathbf{X} = \begin{bmatrix} 1 & t & \frac{1}{2}t^2 \\ 0 & 1 & t \\ 0 & 0 & 1 \end{bmatrix}, \quad \text{(d)}\quad \mathbf{X} = \begin{bmatrix} 1 & \frac{1}{q}[1 - \exp(-qt)] \\ 0 & \exp(-qt) \end{bmatrix},$$

(e) $\mathbf{X}$ is the diagonal matrix with elements $\exp(p_1 t)$, $\exp(p_2 t), \ldots$, $\exp(p_n t)$.

2.4 Differentiating the product $\mathbf{AA}$, we see that $\mathrm{d}(\mathbf{A}^2)/\mathrm{d}t = \mathbf{A}\dot{\mathbf{A}} + \dot{\mathbf{A}}\mathbf{A}$, which equals $2\mathbf{A}\dot{\mathbf{A}}$ if and only if $\mathbf{A}$ and $\dot{\mathbf{A}}$ commute.

Suppose $\mathbf{B}$ is the matrix that satisfies $\dot{\mathbf{B}} = \mathbf{A}$ and $\mathbf{B}(0) = \mathbf{0}$. Then the analogy suggested in the question would give $\mathbf{X}(t, 0) = \exp[\mathbf{B}(t)]$. This would indeed be the correct value for the fundamental matrix if it were true that

$$\frac{\mathrm{d}}{\mathrm{d}t}\exp[\mathbf{B}(t)] = \dot{\mathbf{B}}\exp[\mathbf{B}(t)].$$

Expanding the exponential functions as power series, we see that we require

$$\frac{\mathrm{d}}{\mathrm{d}t}\mathbf{B}^m = m\dot{\mathbf{B}}\mathbf{B}^{m-1} \quad \text{for all integers } m,$$

and, as for the special case with $m = 2$ in the first part of the question, this is true if and only if $\mathbf{B}$ and $\dot{\mathbf{B}}$ commute. This requirement is satisfied when $n = 1$ for all $\mathbf{A}(t)$ and also for general n when $\mathbf{A}$ is a constant matrix, since then $\mathbf{B} = \mathbf{A}t$, which commutes with $\mathbf{A}$.

2.5 The matrix for the adjoint system is

$$\begin{bmatrix} -1 & 0 \\ -1 & 0 \end{bmatrix}$$

and the fundamental matrix has the form

$$\mathbf{Y} = \begin{bmatrix} e^{-t} & 0 \\ e^{-t} - 1 & 1 \end{bmatrix}.$$

With $\mathbf{X}$ given by the solution for Exercise 2.3(a) above, it is easy to verify that $\mathbf{Y}\mathbf{X}^{\mathrm{T}} = \mathbf{I}$.

2.6 For an autonomous system $\mathbf{X}(t - t_0, 0) = \mathbf{X}(t, t_0)$. Hence, from (2.21), it follows that $\mathbf{X}(t, 0)\mathbf{X}(-t, 0) = \mathbf{I}$ and the required result follows from the definition in (2.32).

Expanding the exponentials as power series, we see that we require the binomial theorem to hold for $(\mathbf{A} + \mathbf{B})^m$ for all integers m. With $m = 2$, for example, we have $(\mathbf{A} + \mathbf{B})^2 = \mathbf{A}^2 + \mathbf{AB} + \mathbf{BA} + \mathbf{B}^2 = \mathbf{A}^2 + 2\mathbf{AB} + \mathbf{B}^2$ if and only if $\mathbf{AB} = \mathbf{BA}$ and the matrices $\mathbf{A}$ and $\mathbf{B}$ commute. A similar argument holds for any value of m (or we can use induction) which establishes the second result.

Note that the first part of the question is the special case when $\mathbf{B} = -\mathbf{A}$, and $\mathbf{A}$ and $-\mathbf{A}$ do commute.

2.7 It is easy to show by induction that $\mathbf{A}^m = \mathbf{P}^{-1}\mathbf{B}^m\mathbf{P}$ for any integer m if it is true for $m = 1$. Since $\exp(\mathbf{A}t)$ and $\exp(\mathbf{B}t)$ can be expanded as power series, it follows that they are similar with the same matrix $\mathbf{P}$.

2.8 If we write $y = x_1, \mathrm{d}y/\mathrm{d}t = x_2, \ldots, \mathrm{d}^{n-1}y/\mathrm{d}t^{n-1} = x_n$, then the nth-order differential equation can be replaced by the n first-order equations

$$\dot{x}_1 = x_2, \dot{x}_2 = x_3, \ldots, \dot{x}_{n-1} = x_n,$$
$$\dot{x}_n = -a_0x_1 - a_1x_2 - \cdots - a_{n-1}x_n + f.$$

The matrix of this system is

$$\mathbf{A} = \begin{bmatrix} 0 & 1 & 0 & \cdots & 0 \\ 0 & 0 & 1 & \cdots & 0 \\ \vdots & \vdots & \vdots & \ddots & \vdots \\ -a_0 & -a_1 & -a_2 & \cdots & -a_{n-1} \end{bmatrix}.$$

To find the eigenvalues of $\mathbf{A}$, construct the matrix $\mathbf{A} - p\mathbf{I}$ and perform the column operations of successively adding $p \times$ column 2, $p^2 \times$ column 3, $\ldots$, $p^{n-1} \times$ column n to column 1. The determinant of the transformed matrix is zero if the element in the last row of the first column is zero, which is equivalent to the given algebraic equation for p. If p_i is one of the roots of this equation, the corresponding eigenvector is $(1\ p_i\ p_i^2 \ldots p_i^{n-1})^{\mathrm{T}}$ and the required matrix $\mathbf{P}$ is formed by taking these eigenvectors as its columns. The matrix $\mathbf{\Lambda}$ is the diagonal matrix with elements $p_1, p_2, \ldots, p_n$.

Exercises 3 (p. 44)

3.1 The conditions for controllability are given in §3.3.

(a) In this case, the controllability matrix is given by

$$\mathbf{M} = \begin{bmatrix} 1 & -1 & 0 & 0 \\ 0 & 0 & 0 & 0 \end{bmatrix},$$

which has rank 1, so the system is not completely controllable.

(b) Here

$$\mathbf{M} = \begin{bmatrix} 1 & -1 & 0 & 0 \\ 0 & 0 & 1 & -1 \end{bmatrix},$$

which has rank 2. In case (i) with unbounded controls the system is completely controllable. The eigenvalues of $\mathbf{A}$ are 0 and 1, so that one of them has a positive real part. In case (ii), therefore, the system is not completely controllable.

(c) The determinant of the matrix $\mathbf{M}$ in this case is equal to $-s^3$, so that the system is not completely controllable when $s = 0$. The eigenvalues of $\mathbf{A}$ are each equal to p, so that, for $s \neq 0$, the system is completely controllable in case (i), but only if $p \leq 0$ in case (ii).

3.2 The solution of the state equation at time t_1 has the form

$$x_1^1 = \exp(-t_1)\left(x_1^0 + \int_0^{t_1} \mathrm{e}^t u_1(t)\,\mathrm{d}t\right),$$

and the extreme values of the integral are found by setting $u_1(t) = 1$ or -1, since e^t is positive. The four sets are given by, in order, $x_1^0 = 0$, $x_1^0 = 2$, $x_1^1 = 0$, $x_1^1 = 2$. The answers are as follows:

$$\begin{array}{ll} \mathscr{R}(t_1, 0) & |x_1| \leq 1 - \exp(-t_1), \\ \mathscr{R}(t_1, 2) & -1 + 3\exp(-t_1) \leq x_1 \leq 1 + \exp(-t_1), \\ \mathscr{C}(t_1, 0) & |x_1| \leq \exp(t_1) - 1, \\ \mathscr{C}(t_1, 2) & 1 + \exp(t_1) \leq x_1 \leq 3\exp(t_1) - 1. \end{array}$$

The unions of these sets for all t_1 are

$$\begin{array}{llll} \mathscr{R}(0) & |x_1| < 1, & \mathscr{C}(0) & |x_1| < \infty, \\ \mathscr{R}(2) & -1 < x_1 \leq 2, & \mathscr{C}(2) & 2 \leq x_1 < \infty. \end{array}$$

Consider the point $x_1 = 1 + \exp(-t_1)$. From the above result, $x_1 \in \mathscr{R}(t_1, 2)$. Also, $1 + \exp(-t_1) > 1 + \exp(-t_2)$ when $t_1 < t_2$ so that $x_1 \notin \mathscr{R}(t_2, 2)$. Thus $\mathscr{R}(t_1, 2) \not\subset \mathscr{R}(t_2, 2)$.

3.3 For the time-reversed system,

$$x_1^1 = \mathrm{e}^{t_1}\left(x_1^0 - \int_0^{t_1} \mathrm{e}^{-t} u_1(t)\,\mathrm{d}t\right).$$

The arguments used in Exercise 3.2 when applied to this system lead, for example, to $\mathscr{R}(t_1, 0)$ as the set $|x_1| \leq \mathrm{e}^{t_1} - 1$, which is the same as the set $\mathscr{C}(t_1, 0)$ for the original system.

3.4 The general solution for this system gives $x_1^1 = x_1^0 + t_1 x_2^0 + \int_0^{t_1} u_1 \, dt$, $x_2^1 = x_2^0$. For $\mathscr{C}(t_1, \mathbf{0})$, we set $x_1^1 = x_2^1 = 0$ and find that $x_2^0 = 0$ and $|x_1^0| \leq t_1$. Then $\mathscr{C}(\mathbf{0})$ is the x_1-axis.

For $\mathscr{R}(t_1, \mathbf{x}^0)$ we set $x_1^0 = p$, $x_2^0 = q$, so that $x_2^1 = q$ and

$$p + (q - 1)t_1 \leq x_1^1 \leq p + (q + 1)t_1.$$

3.5 With $\mathbf{A} = \mathbf{0}$ and $\mathbf{B} = \mathbf{I}$, the reachable points at time t_1 are

$$x_1^1 = -1 + \int_0^{t_1} u_1 \, dt, \quad x_2^1 = \int_0^{t_1} u_2 \, dt,$$

so that $|x_1 + 1| \leq t_1$, $|x_2| \leq t_1$, independently of each other. The reachable set at time t_1 is a square of side $2t_1$ and centre $\mathbf{x}^0$. The origin lies in this square whenever $t_1 \geq 1$.

To reach the point $(0, p)^{\mathrm{T}}$ at time $t_1 = 1$ we must choose u_1 and u_2 so that

$$\int_0^1 u_1 \, dt = 1, \quad \int_0^1 u_2 \, dt = p.$$

When $p = 1$, this is only possible if $u_1(t) = 1$ and $u_2(t) = 1$ for $0 < t < 1$. To reach $(0, p)^{\mathrm{T}}$ with $-1 < p < 1$ we must have $u_1(t) = 1$, but there are many possible values for $u_2(t)$. For example, $u_2(t) = p$ or, if $p \geq 0$, $u_2(t) = 1$ when $0 < t < p$ and $u_2(t) = 0$ when $p < t < 1$.

3.6 The fundamental matrix is a special case of (2.17) and the definition (3.6) of the controllable set leads to the given expressions. Without loss of generality we can take α in the range from 0 to π. With $t_1 = \pi$, we see that

$$r = r(\cos^2 \alpha + \sin^2 \alpha) = \int_0^{\pi} u_1(t) \sin(t - \alpha) \, dt,$$

and the maximum value of two is achieved by taking $u_1(t) = -1$ when $0 < t < \alpha$ and $u_1(t) = 1$ when $\alpha < t < \pi$. It is easy to verify that this choice satisfies both equations. By multiplying the control by $\frac{1}{2}r$, we see that all points of the line from $(-2 \cos \alpha, -2 \sin \alpha)$ to $(2 \cos \alpha, 2 \sin \alpha)$ belong to $\mathscr{C}(\pi, \mathbf{0})$. Letting α vary from 0 to π, we see that $\mathscr{C}(\pi, \mathbf{0})$ is the disc $\|\mathbf{x}\| \leq 2$.

A similar choice of u_1 with repeated switches at intervals of π shows that $\mathscr{C}(n\pi, \mathbf{0})$ is the closed disc $\|\mathbf{x}\| \leq 2n$. Since any point in $\mathscr{R}^2$ lies in this disc for sufficiently large n, it follows that $\mathscr{C}(\mathbf{0}) = \mathscr{R}^2$.

Exercises 4 (p. 64)

4.1 Since the system is controllable, the matrix $\mathbf{M}$ has rank n. When $m = 1$ the condition for normality is that the rank of $\mathbf{M}_1$ is n. Since $\mathbf{M}_1 = \mathbf{M}$ the system is normal.

If we apply a similarity transformation, we can reduce $\mathbf{A}$ to diagonal form with elements equal to the distinct eigenvalues λ_i, $i = 1, 2, \ldots, n$. The fundamental matrix is then given by Exercise 2.3(e), and applying the time-optimal control principle we see that the control is given by the sign of an expression like that given for $p_n(t)$.

Suppose that p_n has n changes of sign, that is, has n zeros. Then $\exp(\lambda_1 t)p_n$ also has n zeros and its derivative must have $n - 1$ zeros. The derivative is

$$\exp(\lambda_1 t) \sum_2^n c_i(\lambda_1 - \lambda_i)\exp(-\lambda_i t).$$

Repeating the argument we deduce that

$$c_n \prod_{i=1}^{n-1} (\lambda_i - \lambda_n)\exp(-\lambda_n t)$$

has one zero, which is impossible since all the values of λ_i are distinct and $c_n \neq 0$ in general. Hence the bang–bang optimal control can have at most $n - 1$ switches.

4.2 The state equations for this problem are similar to those in the positioning problem, $\dot{x}_1 = x_2$, $\dot{x}_2 = u_1$, with $|u_1| \leq 10\ \text{m s}^{-2}$, initial state $x_1 = 10^5$ m, $x_2 = 0$, and target the origin. The optimal solution begins with $u_1 = -10\ \text{m s}^{-2}$ and switches to $u_1 = 10\ \text{m s}^{-2}$ after a time $t_2 = 100$ s. The target is reached at time $t_1 = 200$ s. (Apart from the maximum size for the control, this is the same problem as in §4.2(2).)

The maximum speed is reached at the switching time and is equal to $1\ \text{km s}^{-1}$, so the transporter reaches a speed of Mach 3!

4.3 The maximum principle TOP shows that $f = F\ \text{sgn}(\alpha \exp(kt))$ for some constant α, so that we must have either $f = F$ or $f = -F$ throughout, with no switch. If $f = -F$,

$$p = -\frac{F}{k} + \left(P + \frac{F}{k}\right)\exp(-kt),$$

and this is equal to the target value $\frac{1}{2}P$ when

$$t_1 = \frac{1}{k}\ln\left(\frac{F + kP}{F + \frac{1}{2}kP}\right).$$

If $f = F$, the target is reached at time t_2, where

$$t_2 = \frac{1}{k}\ln\left(\frac{kP - F}{\frac{1}{2}kP - F}\right),$$

provided $P > 2F/k$. Since $t_2 > t_1$, there are two extremal solutions, but $f = -F$ gives the optimal time t_1. If $P \leq 2F/k$, the target cannot be reached with the control $f = F$.

To hold the population at the level $\frac{1}{2}P$ we must apply a control equal to $\frac{1}{2}kP$, and this is only possible when $F \geq \frac{1}{2}kP$.

4.4 Since p must increase, the state equation shows we must choose $f = F$. If $P < Q < F/k$, the optimal time is

$$t_1 = \frac{1}{k}\ln\left(\frac{F - kP}{F - kQ}\right).$$

If $Q > F/k$, this desired target state is unattainable. For, as $t_1 \to \infty$, $p \to F/k$ from below if $P < F/k$ and $p < P < Q$ for all positive t if $P > F/k$.

4.5 The maximum principle TOP shows that $u_1 = \operatorname{sgn}((\alpha + \beta)e^{-t})$ for some constants α and β, not both zero. Provided $\alpha + \beta \neq 0$, we must have either $u_1 = 1$ or $u_1 = -1$, without switching. With $u_1 = 1$, the solutions of the state equations are

$$x_1 = -1 + (p + 1)e^t, \quad x_2 = -1 + (q + 1)e^t,$$

and the target is reached in the optimal time $t_1 = -\ln(p + 1)$ provided $p = q$ and $-1 < p \leq 0$. Similarly, if $u_1 = -1$, the optimal time is $t_1 = -\ln(1 - p)$ provided $p = q$ and $0 \leq p < 1$. Hence application of the maximum principle does not ensure the existence of an optimal solution, for no solution exists when $|p| \geq 1$.

The exceptional case, when $\alpha + \beta = 0$, provides no information on u_1. But any choice of u_1 other than the two extreme values enables us to reach only a subset of the reachable set at any time, so the optimal solution is unchanged and the range of values of p and q for which a solution exists is not extended.

4.6 In this problem,

$$\mathbf{A} = \mathbf{0} \quad \text{and} \quad \mathbf{B} = \begin{bmatrix} 1 & 1 \\ 0 & 1 \end{bmatrix}.$$

Hence,

$$\mathbf{M}_1 = \begin{bmatrix} 1 & 0 \\ 0 & 0 \end{bmatrix} \quad \text{and} \quad \mathbf{M}_2 = \begin{bmatrix} 1 & 0 \\ 1 & 0 \end{bmatrix},$$

both of which have rank 1.

5.5 Following the procedure of §5.4, we make the change of variables $x_1 = y_1 - Y_1, x_2 = -1 + y_2 - Y_2, u_1 = v_1 + 1$, and arrive at the same state equations $\dot{x}_1 = x_2 + u_1, \dot{x}_2 = -x_1$ with $-1 \leq u_1 \leq 1$. The initial state is $x_1 = 0$, $x_2 = c - 1$, and the target state is $x_1 = 0$, $x_2 = -1$. As before, the trajectories are arcs of circles with centres at A_1 and A_{-1}, the time taken is equal to the angle subtended at the centre of each circular arc, and switchings take place at intervals of π. With $0 < c < 4$, we start with $u_1 = 1$ and move from the initial point P to a point Q along the circle with centre at A_{-1} and radius c. Then we switch to $u_1 = -1$ and move from Q to the target A_1 along the circle with centre A_1 and radius 2. The position of Q is fixed by the intersection of these two circular arcs and it is easy to show that the triangle $A_1 A_{-1} Q$ has sides equal to 2, 2, and c. The total angle is equal to $\pi - \cos^{-1} \frac{1}{4}c$, and this is the optimal time. The optimal strategy is to do nothing until $t = t_2 = \cos^{-1} \frac{1}{4}c$ ($u_1 = 1$ means that $v_1 = 0$) and to cull the predators at the maximum rate for a time $\pi - 2t_2$.

Since the optimal time using this option is less than π, whereas the alternative, in which extra predators were introduced, required a time greater than π, the culling of the predators is the more efficient option.

5.6 The usual procedure shows that not more than one switch of control is required and since x_2 is initially zero, we must begin with $u_1 = 1$. With this control and the given initial state, we find that $x_2 = 1 - e^{-t}$, $x_1 = 1 - \frac{1}{2}e^{-t} + (X - \frac{1}{2})e^t$. After the switch, the trajectory that reaches the target at time t_1 is given by $x_2 = -1 + \exp(t_1 - t)$, $x_1 = -1 + \frac{1}{2}\exp(t_1 - t) + \frac{1}{2}\exp(t - t_1)$. Equating the two values of x_2 at the switching time t_2, we find that $\exp(t_2) = \frac{1}{2}[\exp(t_1) + 1]$. Equating the two values of x_1 at time t_2 and eliminating t_2, we arrive finally at the result

$$\exp(t_1) = \frac{1 + X + (X^2 + 4X)^{1/2}}{1 - 2X},$$

provided $X < \frac{1}{2}$. As $X \to \frac{1}{2}$, $t_1 \to \infty$.

If $X > \frac{1}{2}$, the value of x_1 in the first section of the trajectory as found above increases monotonically with t, and any subsequent reduction in the number of beetles cannot reverse this increase.

Exercises 6 (p. 97)

6.1 The Hamiltonian has the form $H = -k - \frac{1}{2}u_1^2 + z_1(-x_1 + u_1)$ and the co-state equation is $\dot{z}_1 = z_1$, with solution $z_1 = A\,e^t$. The supremum of H as a function of u_1 is attained when $u_1 = z_1$, and the solution of the state equation is then $x_1 = \frac{1}{2}A\,e^t + B\,e^{-t}$, where A and B are constants. The initial and final states are achieved if we satisfy the equations $B + \frac{1}{2}A = X$, $B\exp(-t_1) + \frac{1}{2}A\exp(t_1) = 0$, where t_1 is the terminal time. The requirement that $H = 0$ along the optimal trajectory yields the condition $AB + k = 0$. Hence, we find that $A^2 - 2AX - 2k = 0$. The negative root of this equation is required (the positive root gives a negative value for t_1), and we find that

$$\exp(t_1) = \frac{X + (X^2 + 2k)^{1/2}}{(2k)^{1/2}}.$$

The optimal control is $u_1 = A\,\mathrm{e}^t$, with $A = X - (X^2 + 2k)^{1/2}$, so the maximum size of control used is $-A\exp(t_1)$, which is equal to $(2k)^{1/2}$.

Substituting this control and the value of t_1 into the integral for J, we find that the optimal cost $J = kt_1 - \frac{1}{2}X^2 + \frac{1}{2}X(X^2 + 2k)^{1/2}$.

6.2 The co-state variable is $z_1 = Ae^t$, as before, but the supremum of H is achieved by taking $u_1 = 1$ when $z_1 > 1$, $u_1 = 0$ when $-1 < z_1 < 1$, and $u_1 = -1$ when $z_1 < -1$. Since z_1 does not change sign, and it is clear that negative values for u_1 are required at least for part of the trajectory, we must have A negative. There are two possibilities: either we employ $u_1 = -1$ throughout, in which case $A \le -1$, or we begin with $u_1 = 0$ and switch to $u_1 = -1$ when $A\,\mathrm{e}^t = -1$.

In the first case, the trajectory is given by $x_1 = -1 + (X + 1)\mathrm{e}^{-t}$ and the terminal time $t_1 = \ln(X + 1)$. The optimal cost is given by $J = (k + 1)\ln(X + 1)$. The condition that $H = 0$ at $t = 0$ gives $A = -(k + 1)/(X + 1)$, so this is the optimal solution provided $A \le -1$, or $k \ge X$.

When $k < X$, we begin with $u_1 = 0$ and $x_1 = X\,\mathrm{e}^{-t}$, $z_1 = -k\,\mathrm{e}^t/X$, making use of the condition $H = 0$. We switch to $u_1 = -1$ when $z_1 = -1$, that is, at $t = t_2$, where $t_2 = \ln(X/k)$. For $t > t_2$ the trajectory is $x_1 = -1 + (X + X/k)\mathrm{e}^{-t}$ and the target is reached at time t_1, where $t_1 = \ln[(k + 1)X/k]$. The optimal cost is $J = kt_1 + t_1 - t_2 = (k + 1)\ln(k + 1) + k\ln(X/k)$.

6.3 The Hamiltonian is $H = -\frac{1}{2}k^2 - \frac{1}{2}u^2 + z_1x_2 + z_2u_1$ and the solutions of the co-state equations are $z_1 = -A$, $z_2 = B + At$. The optimal control $u_1 = z_2$. Hence, $x_2 = Bt + \frac{1}{2}At^2$ and $x_1 = X + \frac{1}{2}Bt^2 + \frac{1}{6}At^3$. If the target is reached at time t_1, these equations give $A = 12X/t_1^3$, $B = -6X/t_1^2$. Setting the value of H to be zero at $t = 0$, we find that $B^2 = k^2$, and since we have already determined that B is negative, $B = -k$. The terminal time $t_1 = (6X/k)^{1/2}$. The optimal control is given by $u_1 = k(2t - t_1)/t_1$ and the optimal cost is

$$J = \tfrac{1}{2}k^2t_1 + \frac{k^2}{2t_1^2}\int_0^{t_1}(2t - t_1)^2\,\mathrm{d}t = \frac{2k}{3}(6kX)^{1/2}.$$

The optimal control varies linearly with time from an initial value of $-k$ to k, so the maximum size of control used is equal to k.

When k is very small, the control employed is also small, but the time taken to reach the target is large. When k is large, it is cost efficient to reach the target quickly, so a large control is used.

6.4 The same methods as used in Exercise 6.3 give three equations for A, B, and t_1, which are now

$$At_1^3 = 6(t_1 + 2), \quad Bt_1^2 = -2(2t_1 + 3), \quad -\tfrac{1}{2}k^2 + \tfrac{1}{2}B^2 - A = 0.$$

Hence the terminal time and optimal cost are found to be

$$t_1 = \frac{(1 + 6k)^{1/2} + 1}{k}, \quad J = \frac{4t_1^2 + 18t_1 + 24}{t_1^3}.$$

The optimal control varies linearly from $-(4t_1 + 6)/t_1^2$ initially to a final value of $(2t_1 + 6)/t_1^2$, and the initial value is the maximum size of control needed.

6.5 The Hamiltonian is $H = -\tfrac{1}{2}k^2 - \tfrac{1}{2}x_1^2 + z_1(-x_1 + 1 - u_1)$ and the co-state equation is $\dot{z}_1 = z_1 + x_1$. The optimal control is $u_1 = -2 \operatorname{sgn} z_1$ and, with $u_1 = 2$, the solution of the state equation is $x_1 = -1 + 2\,\mathrm{e}^{-t}$. The target is reached at time $t_1 = \ln 2$ and the optimal cost $J = \tfrac{1}{2}(k^2 + 1)\ln 2 - \tfrac{1}{4}$.

This is the optimal solution provided we can show that z_1 does not change sign. The initial value of z_1 is given by satisfying the condition that $H = 0$, which gives $z_1^0 = -\tfrac{1}{4}(k^2 + 1)$, and the solution of the co-state equation gives $z_1 = 1 - \mathrm{e}^{-t} - \tfrac{1}{4}(k^2 + 1)\,\mathrm{e}^{t}$. The maximum value of z_1 is equal to $1 - (k^2 + 1)^{1/2}$ so z_1 is negative for all t.

6.6 The Hamiltonian is $H = -1 + z_1(x_1 - x_2) + z_2 u_1$, and the solutions of the co-state equations are $z_1 = A\,\mathrm{e}^{-t}$, $z_2 = B - A\,\mathrm{e}^{-t}$. The control $u_1 = \operatorname{sgn} z_2$, so that no more than one switch can be required. The second state equation shows that, for the given initial and final conditions, we must begin with $u_1 = 1$ and switch to $u_1 = -1$ at some time t_2. Solving for x_2, we find that $x_2 = t$ for $0 < t < t_2$, and $x_2 = 2t_2 - t$ for $t_2 < t < t_1$; the target is reached at time t_1, where $t_1 = 2t_2$. Solving for x_1, we find that $x_1 = t + 1 + (X - 1)\,\mathrm{e}^{t}$ for $0 < t < t_2$, and $x_1 = -t - 1 + 2t_2 + [X - 1 + 2\exp(-t_2)]\mathrm{e}^{t}$ for $t_2 < t < t_1$. Since $x_1 = 0$ when $t = t_1$ and $t_2 = \tfrac{1}{2}t_1$, we arrive at an equation for t_2 in the form

$$\exp(-2t_2) - 2\exp(-t_2) + 1 - X = 0.$$

This equation has an acceptable solution provided $X < 1$ and $\exp(-t_2) = 1 - X^{1/2}$. Hence, $t_1 = -2\ln(1 - X^{1/2})$.

When $X \geq 1$, the number of flies increases even when the maximum number of spiders is introduced, so the plague is not controllable.

Exercises 7 (p. 111)

7.1 The problem in autonomous form becomes, in the standard notation,

$$\dot{x}_0 = u^2, \quad \dot{x}_1 = x_2 u, \quad \dot{x}_2 = 1,$$

and the boundary conditions are $x_0 = x_1 = x_2 = 0$ at $t = 0$ and $x_1 = 1$ at $t = 1$. The Hamiltonian is given by $H = -u^2 + z_1 x_2 u + z_2$, so that $u = \frac{1}{2} z_1 x_2$, and the co-state equations are $\dot{z}_1 = 0$, $\dot{z}_2 = -z_1 u$. Solving the state and co-state equations we find that $z_1 = A$, $z_2 = B - \frac{1}{4}A^2 t^2$, $x_0 = \frac{1}{12}A^2 t^3$, $x_1 = \frac{1}{6}At^3$, $x_2 = t$. From the condition on x_1 at $t = 1$ it follows that $A = 6$ and, since $H = 0$, $B = 0$. The optimal control $u = 3t$ and the optimal cost $J = 3$, as found by elementary methods in Exercise 1.1.

7.2 The co-state equations have solutions of the form $z_1 = A$, $z_2 = B - At$, and the optimal control is $u_1 = \operatorname{sgn} z_2$. The transversality condition shows that $z_1^1 = 0$, so that $A = 0$ and z_2 does not change sign. Hence $u_1 = \pm 1$ with no switch. If $X_2 > 0$, we use $u_1 = -1$ and $x_2 = X_2 - t$. Hence $t_1 = X_2$ and the target is reached at the point where $x_1 = X_1 + \frac{1}{2}X_2^2$.

If $X_2 < 0$, $u_1 = 1$, $t_1 = -X_2$ and $x_1 = X_1 - \frac{1}{2}X_2^2$.

When the target set is $x_1 = -1$, the transversality condition gives $z_2^1 = 0$ and so $z_2 = A(t_1 - t)$. Again there is no switch in u_1, and with $u_1 = -1$ we have $x_2 = X_2 - t$, $x_1 = X_1 + X_2 t - \frac{1}{2}t^2$. The target $x_1 = -1$ is reached at time $t_1 = X_2 + (X_2^2 + 2X_1 + 2)^{1/2}$ and where $x_2 = X_2 - t_1$.

7.3 The Hamiltonian is $H = -k - \frac{1}{2}u_1^2 + z_1 x_2 + z_2 x_3 + z_3 u_1$ and the optimal control is given by $u_1 = z_3$. The co-state equations have the solutions $z_1 = A$, $z_2 = B - At$, $z_3 = C - Bt + \frac{1}{2}At^2$. The transversality condition shows that $z_1^1 = 0$, $z_2^1 = 0$, so that $A = 0$, $B = 0$, and $u_1 = z_3 = C$. Thus $x_3 = X_3 + Ct$ and $Ct_1 = -X_3$. The zero value of the Hamiltonian along the trajectory shows that $C^2 = 2k$ and, since C must be negative to give a positive terminal time, $t_1 = X_3/(2k)^{1/2}$. The optimal cost is $J = (k + \frac{1}{2}C^2)t_1 = (2k)^{1/2}X_3$ and the final state of the system is

$$x_1^1 = X_1 + \frac{X_2 X_3}{(2k)^{1/2}} + \frac{X_3^3}{6k}, \quad x_2^1 = X_2 + \frac{X_3^2}{2(2k)^{1/2}}, \quad x_3^1 = 0.$$

7.4 As in Exercise 7.2, $z_1 = 0$, $z_2 = 0$, $z_3 = C$, but the supremum of H requires $u_1 = 1$ when $z_3 > 1$, $u_1 = 0$ when $-1 < z_3 < 1$, and $u_1 = -1$ when $z_3 < -1$. Since z_3 is constant there are no switches. From $H = 0$ we find that either $C = k + 1 > 1$ and $u_1 = 1$, or $C = -k - 1 < -1$ and $u_1 = -1$; $u_1 = 0$ is not possible. Since X_3 is positive, we choose

$u_1 = -1$, and the solution for x_3 is $x_3 = X_3 - t$. Thus $t_1 = X_3$, the optimal cost is $J = (k+1)X_3$, and the terminal state is given by

$$x_1^1 = X_1 + X_2X_3 + \tfrac{1}{3}X_3^3, \quad x_2^1 = X_2 + \tfrac{1}{2}X_3^2, \quad x_3^1 = 0.$$

7.5 The solutions of the co-state equations that satisfy the transversality condition $z_1^1 = 0$ have the form $z_1 = A\sin(t - t_1)$, $z_2 = A\cos(t - t_1)$, and the supremum of H requires that u_1 should be $+1, 0, -1$ according as z_1 is greater than $+1$, between -1 and $+1$, or less than -1. If we employ $u_1 = -1$ throughout, the trajectory is an arc of the circle with centre at $(-1, 0)$ through the initial point. The optimal time is $t_1 = \tan^{-1} X$ and the optimal cost $J = 2\tan^{-1} X$. The value of z_2^1 is fixed by the condition $H = 0$ and we find that $A = -2/(1 + X^2)^{1/2}$. From this result we can show that $z_2 < -1$, as required if there is to be no switch in the value of u_1, provided $X < 1$. If we use the control $u_1 = 0$ throughout, the trajectory is an arc of a circle with centre at the origin, and $t_1 = \tfrac{1}{2}\pi$. The optimal cost $J = \tfrac{1}{2}\pi$. The value of z_2 is found to be equal to $-X^{-1}\sin t$, so that for the whole of the trajectory z_2 lies between -1 and 0, as required, provided $X > 1$. The possibilities that we should start with $u_1 = 0$ and switch to $u_1 = -1$, or the other way round, must also be examined, but they do not provide a lower cost for any value of X, and the solutions found are the optimal ones.

Note that when $X = 1$, both $u_1 = -1$ and $u_1 = 0$ are successful controls, and both involve the same optimal cost $\tfrac{1}{2}\pi$. The first choice reaches the target in time $\tfrac{1}{4}\pi$ and the second in time $\tfrac{1}{2}\pi$. In fact, we can use $u_1 = 0$ for a time $t_2 < \tfrac{1}{2}\pi$ and $u_1 = -1$ from t_2 to $\tfrac{1}{4}\pi + \tfrac{1}{2}t_2$ to reach the target with the same optimal cost.

7.6 From the Hamiltonian we find that $u_1 = \operatorname{sgn} z_1$, $u_2 = \operatorname{sgn} z_2$, and that z_1 and z_2 are both constants. Hence the trajectory meets the target at time t_1 if

$$r\cos t_1 = \pm t_1, \quad r\sin t_1 = \pm t_1,$$

which only have solutions for $r = t_1\sqrt{2}$ and when $t_1 = \tfrac{1}{4}\pi + n\pi$, for integral values of n. When $r = \pi/\sqrt{8}$, $t_1 = \tfrac{1}{4}\pi$ and $u_1 = 1$, $u_2 = 1$.

For values of r not included in this set, we consider the possibility that either z_1 or z_2 is zero. If $z_1 = 0$, the control u_1 is not determined. With $u_2 = 1$, we find that $x_2^1 = t_1 = r\sin t_1$, with $t_1 = \tfrac{1}{2}\pi$ when $r = \tfrac{1}{2}\pi$. The target value of x_1 at this time is $x_1^1 = 0$, so that we have an acceptable solution, provided we can choose the control u_1 to satisfy the condition

$$\int_0^{\pi/2} u_1 \, dt = 0,$$

which is possible in many different ways.

7.7 If the force is denoted by x_3, the second state equation equates the force to the rate of change of velocity, and the third equation introduces the controllable rate of increase of the force. The solutions of the co-state equations are $z_1 = A$, $z_2 = B - At$, $z_3 = C - Bt + \frac{1}{2}At^2$. The transversality condition, which is present because the terminal value of x_3 is not specified, is $z_3^1 = 0$. Since z_3 is a quadratic function of t, with one root at the terminal time t_1, there can be at most one switch of the control $u_1 = \operatorname{sgn} z_3$. It is clear that, for the given initial and final states, x_3 must take both positive and negative values, and one switch is required; it is also clear that we must begin with $u_1 = -1$. If we suppose that the switch occurs at $t = t_2$, the solutions of the state equations show that, at time t_2,

$$x_3 = -t_2, \quad x_2 = -\tfrac{1}{2}t_2^2, \quad x_1 = X - \tfrac{1}{6}t_2^3.$$

For $t > t_2$ we use $u_1 = 1$, and reach the target at time t_1 if

$$\tfrac{1}{2}t_1^2 - 2t_1t_2 + t_2^2 = 0, \quad X + \tfrac{1}{6}t_1^3 - t_1^2t_2 + t_1t_2^2 - \tfrac{1}{3}t_2^3 = 0,$$

and from these equations we find that

$$t_2 = \frac{\sqrt{2} - 1}{\sqrt{2}}t_1, \quad t_1^3 = \frac{6\sqrt{2}}{\sqrt{2} - 1}X.$$

The final value of x_3 is not zero in this solution, and, if we wish to end with all three components equal to zero, we must allow the control to switch twice (the condition $z_3^1 = 0$ is no longer applicable since the target is fixed). With $u_1 = -1$ for $0 < t < t_2$ and for $t_3 < t < t_1$, we find that

$$x_3 = -t_2, x_2 = -\tfrac{1}{2}t_2^2, x_1 = X - \tfrac{1}{2}t_2^3, \quad \text{at } t = t_2,$$
$$x_3 = t_1 - t_3, x_2 = -\tfrac{1}{2}(t_1 - t_3)^2, x_1 = \tfrac{1}{6}(t_1 - t_3)^3, \quad \text{at } t = t_3.$$

From t_2 to t_3 we use $u_1 = 1$, and the solution for this interval involves three arbitrary constants. Matching the trajectories at t_2 and t_3 gives six equations for these three constants and the three times. The solution is

$$t_2 = \tfrac{1}{4}t_1, \quad t_3 = \tfrac{3}{4}t_1, \quad t_1 = (32X)^{1/3}.$$

The optimal time without the requirement that x_3 be brought to zero was approximately equal to $2.736X^{1/3}$, while with this extra condition it is increased slightly to $2.773X^{1/3}$.

7.8 To minimize f we must choose u to give the supremum of

$$H = -\mu kx(1-u) - \lambda kxu + zkux(1-\lambda),$$

with $0 \le u \le 1$, so that $u = 1$ when $(1-\lambda)z > \lambda - \mu$ and $u = 0$ otherwise. The co-state equation is $\dot{z} = \mu k + (\lambda - \mu)ku - (1-\lambda)kuz$ and, since the target is free, $z(T) = -\mu$ from the transversality condition with a terminal cost. The condition that determines u shows that $u = 0$ at $t = T$. For $t < T$ we then see that z is increasing so that $z(t) < -\mu$ and we must use $u = 0$ for all t. Hence, $z = \mu k(t - T) - \mu$ and $\dot{x} = 0$, so that $x = X$. The minimum value of $f(T)$ is therefore equal to $\mu X + \int_0^T \mu kx \, dt = \mu X(1 + kT)$.

To minimize $-e$, the Hamiltonian is $H = (1-u)kx + zkux(1-\lambda)$, so that $u = 1$ when $(1-\lambda)z > 1$ and $u = 0$ when $(1-\lambda)z < 1$. The co-state equation is $\dot{z} = -k(1-u) - zku(1-\lambda)$. The transversality condition is $z(T) = 1$, so that we must end with $u = 0$, but we may either use this control throughout or begin with $u = 1$ and switch to $u = 0$ at some time t_1. In the first case, $\dot{z} = -k$ and $\dot{x} = 0$, so that $z = 1 - k(t - T)$ and $x = X$. The optimal value of $e = (1 + kT)X$. This is the solution provided $(1-\lambda)z < 1$ throughout, which means that $kT < \lambda/(1-\lambda)$. If this condition is not satisfied, we switch controls when $(1-\lambda)z = 1$ at time t_1, so that $k(T - t_1) = \lambda/(1-\lambda)$. For $0 < t < t_1$ we use $u = 1$ and $x = X \exp[(1-\lambda)kt]$. For $t > t_1$, $u = 0$ and $x = X \exp[(1-\lambda)kt_1]$. The maximum value of $e(T)$ is given by

$$e(T) = X \exp[(1-\lambda)kt_1] + \int_{t_1}^{T} kX \exp[(1-\lambda)kt_1] \, dt,$$

which simplifies to give

$$e(T) = \frac{X}{1-\lambda} \exp[(1-\lambda)kt_1].$$

7.9 We introduce a new variable x_3, satisfying the equation $\dot{x}_3 = 1$, $x_3^0 = 0$. The Hamiltonian is $H = -1 + z_1x_2 + z_2(x_3 + u_1) + z_3$. The optimal control is $u_1 = \operatorname{sgn} z_2$, and the co-state equations show that z_2 is a linear function of t, so that there can be at most one switch of the control. It is clear that with the given initial and final states, we must begin with $u_1 = -1$ and end with $u_1 = 1$. Hence $x_2 = \frac{1}{2}t^2 - t$, $x_1 = X + \frac{1}{6}t^3 - \frac{1}{2}t^2$ for $t < t_2$, the switching time, and $x_2 = \frac{1}{2}(t - t_1)^2 + (1 + t_1)(t - t_1)$, $x_1 = \frac{1}{6}(t - t_1)^3 + \frac{1}{2}(1 + t_1)(t - t_1)^2$ for $t > t_2$. Equating the two values of x_2 at $t = t_2$, we find that $2t_2 = t_1 + \frac{1}{2}t_1^2$ and, since $t_2 \le t_1$, $t_1 \le 2$. The two values of x_1 at $t = t_2$ give the equation

$$48X = t_1^2(12 + 4t_1 - 3t_1^2).$$

The right-hand side has a maximum value of 32 when $t_1 = 2$. For $X < \frac{2}{3}$ there are two positive roots, but the smaller one is, of course, the optimal solution.

For $X > \frac{2}{3}$, x_2 is positive and increasing when $t > 2$ even when $u_1 = -1$ and it is impossible to reach the target.

7.10 Introduce the variable x_3 to replace t, so that the state equations become

$$\dot{x}_1 = x_2, \quad \dot{x}_2 = -x_1 + 2\cos x_3 + u_1, \quad \dot{x}_3 = 1,$$

and the Hamiltonian is

$$H = -u_1^2 + z_1 x_2 + z_2(-x_1 + 2\cos x_3 + u_1) + z_3.$$

The solutions of the first two co-state equations can be taken as $z_1 = A\sin(t - \alpha)$, $z_2 = A\cos(t - \alpha)$, and the optimal control is given by $u_1 = \frac{1}{2}z_2$. The initial and final values of the state variables are $x_1^0 = 0$, $x_2^0 = 0$, $x_3^0 = 0$, $x_1^1 = 0$, $x_2^1 = 0$, $x_3^1 = t_1$. The solutions of the state equations and these initial and final conditions yield two equations for the constants A and α, which can be reduced to the forms

$A[t_1\sin(t_1 - \alpha) + \sin t_1 \sin\alpha] = -4t_1 \sin t_1$, $(t_1^2 - \sin^2 t_1)\sin\alpha = 0$. Hence, $\alpha = 0$ and $A = -4$, or $\alpha = \pi$ and $A = 4$. In both cases, $u_1 = -2\cos t$. The system remains in its equilibrium position, and the cost is given by

$$J = \int_0^{t_1} 4\cos^2 t\,dt = 2t_1 + \sin 2t_1.$$

Exercises 8 (p. 130)

8.1 The condition $H = 0$ at $t = t_0$ gives the equation

$$\frac{b^2}{2r}(z_1^0)^2 + a x_1^0 z_1^0 - \tfrac{1}{2}q(x_1^0)^2 = 0,$$

and from (8.9) we find that

$$z_1^0 = B_1 + B_2 = \frac{r}{b^2}[\lambda(A_1 - A_2) - a x_1^0].$$

From these two equations, we obtain the result $A_1 - A_2 = \pm x_1^0$ for any value of T. The values of A_1 and A_2 from (8.10) then give that $\lambda = |fb^2r^{-1} - a|$, or $q = f(fb^2r^{-1} - 2a)$. The free-time problem only has a solution when this condition is satisfied, and then there is a solution for all times T.

8.2 The state and co-state equations are

$$\dot{x}_1 = ax_1 + b^2r^{-1}z_1, \quad \dot{z}_1 = qx_1 - az_1,$$

with $x_1^1 = 0$. The solutions are as given in (8.8) and (8.9), and with the new terminal value of x_1 we find that

$$A_1 = -\frac{x_1^0 \exp(-\lambda T)}{2\sinh \lambda T}, \quad A_2 = \frac{x_1^0 \exp(\lambda T)}{2 \sinh \lambda T},$$

$$B_1 + B_2 = -\frac{r}{b^2}(x_1^0)^2(a + \lambda \coth \lambda T).$$

The optimal cost is given by $J = -\frac{1}{2}z_1^0 x_1^0$ and $z_1^0 = B_1 + B_2$, so that

$$J = \frac{r}{2b^2}(x_1^0)^2(a + \lambda \coth \lambda T).$$

We see that $J \to \infty$ as $T \to 0$ and that $J \to (r/2b^2)(x_1^0)^2(a + \lambda)$ as $T \to \infty$. There is no optimal solution for the free-time problem, since the minimum cost is not attained at any finite value of T.

8.3 With the indicated substitution into the state and co-state equations, we arrive at the equation $\dot{L} = qL^2 + 2aL - b^2r^{-1}$. The solution of this equation is $L = (\lambda/q)\tanh(\beta + \lambda t_1 - \lambda t) - (a/q)$, with $\lambda \tanh \beta = a$ and $\lambda = (a^2 + qb^2r^{-1})^{1/2}$. At $t = t_0$, we find after simplification that

$$L(t_0, t_1) = \frac{b^2r^{-1} \tanh \lambda T}{\lambda + a \tanh \lambda T},$$

where $T = t_1 - t_0$. The optimal cost $J = -\frac{1}{2}z_1^0 x_1^0$, so that

$$J = \frac{\frac{1}{2}(x_1^0)^2}{L(t_0, t_1)} = \frac{r}{2b^2}(x_1^0)^2(a + \lambda \coth \lambda T),$$

as found in Exercise 8.2.

8.4 Solving the state and co-state equations in the usual way, we find that

$$x_1 = X\frac{\sinh(t_1 - t)}{\sinh t_1}, \quad z_1 = u_1 = -X\frac{\cosh(t_1 - t)}{\sinh t_1},$$

and setting $H = 0$ at t_1 we find that $\sinh t_1 = X/c$. The optimal cost is given by $J = -\frac{1}{2}z_1^0 X + \frac{1}{2}c^2 t_1 = \frac{1}{2}c^2 \sinh^{-1}(X/c) + \frac{1}{2}X(c^2 + X^2)^{1/2}$.

For the free-target problem with a fixed terminal time, we have the transversality condition $z_1^1 = -x_1^1$, and the solution is $x_1 = Xe^{-t}$, $z_1 = -Xe^{-t}$. The optimal cost is given by $J = \frac{1}{2}(X^2 + c^2 t_1)$. For the free-time problem, we see at once that the minimum cost is $J = \frac{1}{2}X^2$ with $t_1 = 0$.

8.5 We can solve this problem using the method of §8.4, setting $\mathbf{x} = -\mathbf{L}(t, t_1)\mathbf{z}$. If we write

$$\mathbf{L} = \begin{bmatrix} l_1 & l_2 \\ l_2 & l_3 \end{bmatrix},$$

the equations for the components of $\mathbf{L}$ are

$$\dot{l}_1 = 2l_2, \quad \dot{l}_2 = l_3 - l_1, \quad \dot{l}_3 = -2l_2 - 1,$$

and, since $\mathbf{L}(t_1, t_1) = \mathbf{0}$, the solutions of these equations have the form

$$l_{1,3} = -\tfrac{1}{2}(t - t_1) \pm \tfrac{1}{4}\sin 2(t - t_1), \quad l_2 = -\tfrac{1}{4} + \tfrac{1}{4}\cos 2(t - t_1).$$

To find the optimal cost, we use the result (8.57), so we must evaluate the inverse of $\mathbf{L}(0, t_1)$. When this is done, and the given initial conditions are substituted into (8.57), we obtain

$$J = \frac{X_1^2(t_1 + sc) + 2X_1X_2s^2 + X_2^2(t_1 - sc)}{t_1^2 - s^2},$$

where we have written $\sin t_1 = s$, $\cos t_1 = c$.

8.6 With the same notation as in the solution of Exercise 8.5, the equations for the components of $\mathbf{L}$ are now

$$\dot{l}_1 = 2l_2 + l_1^2, \quad \dot{l}_2 = l_3 - l_1 + l_1 l_2, \quad \dot{l}_3 = -2l_2 - 1 + l_2^2.$$

To find the limiting solution as $t_1 \to \infty$, we can set the right-hand sides of these equations to zero. The solution is

$$\mathbf{L}(0, \infty) = \begin{bmatrix} \sqrt{2}(\sqrt{2} - 1)^{1/2} & 1 - \sqrt{2} \\ 1 - \sqrt{2} & 2(\sqrt{2} - 1)^{1/2} \end{bmatrix},$$

where the ambiguity in sign in the solution is removed because the matrix is known to be positive definite. The matrix is easy to invert since its determinant is equal to unity. For the given initial conditions, we then have the limiting optimal cost in the form

$$J = X_1^2(\sqrt{2} - 1)^{1/2} + X_1X_2(\sqrt{2} - 1) + \tfrac{1}{2}X_2^2\sqrt{2}(\sqrt{2} - 1)^{1/2}.$$

8.7 In this problem, $H = -\tfrac{1}{2}(u_1 - u_2)^2 + z_1(x_2 + u_1) + z_2(x_1 + u_2)$ and the supremum is given by $z_1 = u_1 - u_2$, $z_2 = u_2 - u_1$. Hence, $z_1 + z_2 = 0$ and $u_1 - u_2 = z_1$. The solutions of the co-state equations are $z_1 = Ae^t + Be^{-t}$, $z_2 = -Ae^t + Be^{-t}$, so that $B = 0$ and $u_1 - u_2 = Ae^t$. Solving the equation for $x_1 - x_2$, we find that $x_1 - x_2 = \tfrac{1}{2}Ae^t + Ce^{-t}$, and from the initial and final conditions on $x_1 - x_2$, we find that $X_1 - X_2 = \tfrac{1}{2}A[1 - \exp(2T)]$. The optimal cost is given by

$$J = \int_0^T \tfrac{1}{2}A^2 \exp(2t)\,\mathrm{d}t = \frac{(X_1 - X_2)^2}{\exp(2T) - 1}.$$

Thus the difference between the controls is fixed. The control u_1, say, must then be chosen to ensure that x_1 reaches its target value, but it is easy to verify that this can be done in many ways, as u_1 is only required to satisfy an integral constraint. Note that if $X_1 = X_2$, the optimal cost is zero, since then the target can be reached with $u_1 = u_2$ and the trajectory lies along the line $x_1 = x_2$.

Exercises 9 (p. 150)

9.1 For the problem specified in §5.3, the Hamiltonian is

$$H = -1 + z_1 x_2 + z_2(-x_1 + u_1),$$

so that the co-state equations are $\dot{z}_1 = z_2$, $\dot{z}_2 = -z_1$, with solutions $z_1 = A\sin(t - \alpha)$, $z_2 = -A\cos(t - \alpha)$. The supremum of H is attained when $u_1 = \operatorname{sgn} z_2$. Hence, u_1 takes the values of $+1$ or -1, switching at time intervals equal to π. This is the same conclusion as was reached in §5.3 by applying the maximum principle TOP developed in Chapter 4, and the rest of solution follows in exactly the same way as described there.

9.2 For the problem in §6.4, there are discontinuities in the control at $t = t_2$ and at $t = t_3$. For $0 < t < t_2$, $u_1 = -1$ and $H = -k - 1 + z_1 x_2 - z_2$; for $t_2 < t < t_3$, $u_1 = 0$ and $H = -k + z_1 x_2$; for $t_3 < t < t_1$, $u_1 = 1$ and $H = -k - 1 + z_1 x_2 + z_2$. But, from (6.71), $z_2 = -1$ at $t = t_2$ and $z_2 = 1$ at $t = t_3$. Both z_1 and x_2 are continuous functions of t (they both satisfy first-order differential equations), so H is continuous at the switching times t_2 and t_3.

9.3 The optimal trajectory and the perturbations are sketched in Fig. S.1, where x_0 is the state variable associated with the cost; in this case x_0 satisfies the equation $\dot{x}_0 = 1$.

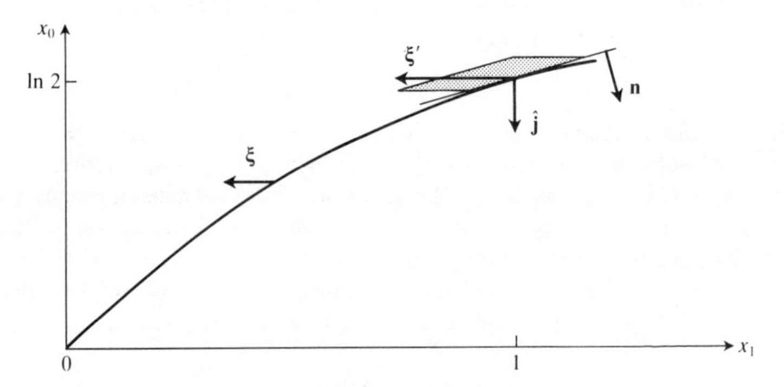

Fig. S.1. Optimal trajectory and perturbations.

The optimal control is $u_1 = 1$ and the optimal trajectory is $x_1 = e^t - 1$. The target is reached at time $t_1 = \ln 2$ and the optimal cost is $J = \ln 2$. If we perturb the control from $\tau - \varepsilon$ to τ, changing it to v, the perturbation at time τ will be $\xi_1 = \varepsilon(v - 1)$, $\xi_0 = 0$. Using the optimal control from $t = \tau$ to $t = t_1$, the perturbation at t_1 is equal to $\xi_1 = \varepsilon(v - 1)\exp(t_1 - \tau)$, $\xi_0 = 0$. The perturbation cone in this case is the ray pointing in the negative x_1 direction; since $u_1 \leq 1$, $v - 1$ is negative. Now consider a perturbation δ in the terminal time. This produces a perturbation $\xi_0 = \delta$, $\xi_1 = 2\delta$, and both signs for δ are possible. Combining the two types of perturbation, the perturbation cone forms a semicircle with centre at the terminal point $x_0 = \ln 2$, $x_1 = 1$ and the outward normal to the diameter of this semicircle points in the direction $(-2, 1)$. The diameter separates the perturbation cone from the downward pointing $\hat{\mathbf{j}} = (-1, 0)$. The co-state equations are adjoint to the perturbed state equations, so they are given by $\dot{z}_1 = -z_1$, $\dot{z}_0 = 0$. At time t_1, we have $z_1 = \frac{1}{2}$, $z_0 = -1$ so that, at time τ, $z_1 = \frac{1}{2}\exp(t_1 - \tau)$, $z_0 = -1$. Also, $\hat{\mathbf{z}}^{\mathrm{T}}\hat{\boldsymbol{\xi}} = \frac{1}{2}\varepsilon(v - 1) < 0$ for all t between τ and t_1, and therefore the optimal control gives H its supremum over all admissible controls.

Exercises 10 (p. 169)

10.1 It is clear that we must end with $u_1 = 1$, and the solution of the co-state equation for z_2 and the condition $H = 0$ at the terminal time t_1 show that

$$z_2 = \frac{k + 1}{1 - c} + A(t_1 - t).$$

Since z_2 is a linear function of t, and we switch controls when z_2 passes through the values $+1$ and -1, the optimal strategy is either to use the sequence $\{0, 1\}$, provided $z_2(0) \geq -1$, or the sequence $\{-1, 0, 1\}$ when $z_2(0) < -1$. In the first case, suppose the switch takes place at $t = t_2$. Then, for $t_2 < t < t_1$, we can solve the state equations with $u_1 = 1$, and find that

$$x_2(t_2) = (1 - c)(t_2 - t_1), \quad x_1(t_2) = \tfrac{1}{2}(1 - c)(t_2 - t_1)^2,$$

$$z_2(t_2) = \frac{k + 1}{1 - c} + A(t_1 - t_2) = 1.$$

For $0 < t < t_2$ we use $u_1 = 0$ and the solution at $t = 0$ is

$$x_2(0) = t_2 - (1 - c)t_1 = 0,$$

$$x_1(0) = -\tfrac{1}{2}ct_2^2 - (1 - c)t_2(t_2 - t_1) + \tfrac{1}{2}(1 - c)(t_2 - t_1)^2 = h,$$

$$z_2(0) = \frac{k + 1}{1 - c} + At_1 \geq -1.$$

From these equations we find that

$$t_1^2 = \frac{2h}{c(1-c)}, \quad t_2 = (1-c)t_1, \quad J = kt_1 + t_1 - t_2 = (k+c)t_1.$$

Also, the condition on $z_2(0)$ becomes $-k/c \geq -1$, or $k \leq c$.

When $k > c$, we must use the alternative sequence of controls, with $u_1 = -1$ from 0 to t_3, $u_1 = 0$ from t_3 to t_2, and $u_1 = 1$ from t_2 to t_1, with $z_2 = -1$ at time t_3 and $z_2 = 1$ at time t_2. Proceeding as before, we find that

$$t_2 = \frac{2+k+c}{2(1+k)}(1-c)t_1, \quad t_3 = \frac{k-c}{2(1+k)}(1-c)t_1,$$

$$t_1^2 = \frac{4h(1+k)^2}{(1-c^2)(k^2+2k+c^2)}.$$

The optimal cost is given by

$$J = kt_1 + t_1 - t_2 + t_3 = 2\left(\frac{h(k^2+2k+c^2)}{1-c^2}\right)^{1/2}.$$

10.2 The transversality condition shows that $z_1^1 = 0$, and the appropriate solutions of the co-state equations have the form $z_1 = A\sin(t - t_1)$, $z_2 = -A\cos(t - t_1)$. The optimal controls are given by $u_1 = z_1$, $u_2 = z_2$, and the solutions of the state equations satisfying the initial conditions are

$$x_1 = At\sin(t - t_1) - X\sin t, \quad x_2 = -At\cos(t - t_1) + X\cos t.$$

Since $x_2(t_1) = 0$, $At_1 = X\cos t_1$; the condition that $H = 0$ at $t = 0$ gives

$$-\tfrac{1}{2} + \tfrac{1}{2}A^2 + AX\sin t_1 = 0,$$

so that $t_1^2 - 2X^2 t_1 \sin t_1 \cos t_1 = X^2\cos^2 t_1$. The optimal cost is given by

$$J = \tfrac{1}{2}t_1 + \tfrac{1}{2}A^2 t_1 = \frac{t_1^2 + X^2\cos^2 t_1}{2t_1}.$$

10.3 From the supremum of the Hamiltonian we find that $u_1 = 1$ when $z_1 > 1$ and $u_1 = -1$ when $z_1 < 1$. The zero value of H at $t = t_1$ gives that $z_1 = 1$ then, and from the state equations we deduce that we must approach the target using $u_1 = 1$. The zero value of H at $t = 0$ gives $z_1 = 0$ if $u_1 = -1$ and $z_1 = 2/(2+c)$ if $u_1 = 1$. The second option is impossible since $c > 0$. From these conditions and the co-state equations, it follows that $z_1 = -A \sin t$. If we apply the control $u_1 = -1$ until $t = \pi + \alpha$, switch to $u_1 = 1$ for t between $\pi + \alpha$ and $2\pi - \alpha$, and revert to $u_1 = -1$ for t between $2\pi - \alpha$ and 2π, we reach the point $x_1 = 0$, $x_2 = 1 + c - 4 \cos \alpha$. The cost incurred during this sequence is equal to $2(\pi - 2\alpha)$. The target is reached if we choose $\cos \alpha = c/4$ and the cost is then given by $J_1 = 4 \sin^{-1}(c/4)$. By repeating the cycle of controls n times, we find that there is a solution satisfying the PMP with a cost given by $J_n = 4n \sin^{-1}(c/4n)$. The infimum of these costs is equal to c, but it is never attained. Thus there is no optimal solution.

Note that in this problem the control can continue for a time greater than π before it has to switch. Two switches have to occur in an interval of 2π.

10.4 Since there is no restriction on the initial and final values of x_3, we apply transversality conditions at $t = 0$ and at $t = t_1$ (see §9.4 for the extension of the transversality condition to cover unspecified initial states). These conditions are that $z_3^0 = 0$ and $z_3^1 = 0$. It follows that we begin and end with $u_1 = 0$. Let $x_3^0 = -p$; then the zero value of H gives that $z_2^0 = -kh/p$ and $z_2^1 = kh/p$. From the co-state equations it follows that $z_3 = (kh/pt_1)t(t_1 - t)$. The control switches to $u_1 = 1$ when $t = t_2$, and switches back to $u_1 = 0$ when $t = t_1 - t_2$, at which times $z_3 = 1$, so that $pt_1 = kht_2(t_1 - t_2)$. The trajectory is clearly symmetrical about $t = \frac{1}{2}t_1$ and we need only consider the interval $0 < t < \frac{1}{2}t_1$. The solution of the third state equation is $x_3 = -p$ for $0 < t < t_2$ and $x_3 = -p + h(t - t_2)$ for $t_2 < t < \frac{1}{2}t_1$. The symmetry shows that $x_3(\frac{1}{2}t_1) = 0$, so that $p = h(\frac{1}{2}t_1 - t_2)$ and, with the condition arrived at from the value of z_3 at the switch, we find that $t_1(\frac{1}{2}t_1 - t_2) = kt_2(t_1 - t_2)$.

The second equation connecting t_2 and t_1 comes from the value of x_1 at t_1, which, by the symmetry, must equal $\frac{1}{2}X$. Solving the first and second state equations with the given initial conditions to find the value of $x_1(\frac{1}{2}t_1)$, we obtain the equation

$$X = \tfrac{1}{3}h(\tfrac{1}{2}t_1 - t_2)(\tfrac{1}{2}t_1^2 + t_1 t_2 - t_2^2).$$

These two equations suffice to determine t_1 and t_2. The optimal cost is

$$J = kht_1 + h(t_1 - 2t_2).$$

10.5 From the discussion in §10.4 concerning this problem, it follows that for all initial states one or other of the following control sequences must be used: (a) $u_1 = 0$, $u_2 = -1$ followed by $u_1 = -1$, $u_2 = 0$; (b) $u_1 = 1$, $u_2 = 1$ followed by $u_1 = -1$, $u_2 = 0$; (c) $u_1 = 0$, $u_2 = -1$ followed by $u_1 = 1$, $u_2 = 1$. The solutions of the state equations can be found for arbitrary u_1 and u_2 and for arbitrary X_1 and X_2. These solutions have the form

$$x_1(t) = \mathrm{e}^{-t}\left(X_1 + \int_0^t \mathrm{e}^{\tau} u_1(\tau)\,\mathrm{d}\tau\right), \quad x_2(t) = X_2 + \int_0^t u_2(\tau)\,\mathrm{d}\tau,$$

so that the target is reached if $X_1 = -\int_0^{t_1} \mathrm{e}^t u_1\,\mathrm{d}t$, $X_2 = -\int_0^{t_1} u_2\,\mathrm{d}t$.

The sequence of controls (a) is that used in the text for $X_1 > 0$, $X_2 > 0$. For case (b) we find that $X_2 < 0$, $X_1 > 1 - \exp(-X_2)$. Case (c) covers the remainder of the plane, that is, $X_1 < 0$, $\exp(-X_2) > 1 - X_1$. These results follow by substituting into the integrals for X_1 and X_2 the appropriate sequence of controls, with the switch occurring at some time t_2, where $0 < t_2 < t_1$. Initial states lying on the boundaries of the three regions of the (x_1, x_2) plane are controlled by one of the three pairs of controls acting alone. Thus, if $u_1 = -1$, $u_2 = 0$, the initial state is $X_1 > 0$, $X_2 = 0$; if $u_1 = 0$, $u_2 = -1$, it is $X_1 = 0$, $X_2 > 0$; if $u_1 = 1$, $u_2 = 1$, then $X_2 < 0$, $X_1 = 1 - \exp(-X_2)$.

10.6 The co-state variables z_1 and z_2 are constants, and we can write $z_1 = A\cos\alpha$, $z_2 = A\sin\alpha$. The supremum of H then requires that $u_1\cos\alpha + u_2\sin\alpha$ have its maximum value subject to $u_1^2 + u_2^2 \le 1$. Hence we can take $u_1 = -\cos\alpha$, $u_2 = -\sin\alpha$ with $0 < \alpha < \pi$, since we clearly need u_2 to be negative. Then

$$x_1 = X - t(V + \cos\alpha), \quad x_2 = 1 - t\sin\alpha,$$

and the target is reached at time $t_1 = 1/\sin\alpha$, provided that

$$X = \frac{V + \cos\alpha}{\sin\alpha}.$$

If $0 < V < 1$, a value of α exists for all values of X. If $V > 1$, the right-hand side has a minimum value of $(V^2 - 1)^{1/2}$ when $\cos\alpha = -1/V$, so that a solution is only possible when $X \ge (V^2 - 1)^{1/2}$.

When the landing point is not specified, the target is the line $x_2 = 0$ and the transversality condition requires $z_1^1 = 0$. Thus, $u_2 = -1$, $u_1 = 0$, and $\alpha = \pi$. Then $t_1 = 1$ for all values of X, and the bank is reached at the point where $x_1 = X - V$.

10.7 Without the restriction on the value of x_2, the solution can be found in the usual manner. The control $u_1 = z_2 = -k - At$, from the co-state equations and the condition $H = 0$ at $t = 0$. Solving the state equations with the given initial and final conditions gives $At_1 = -2k$ and $kt_1^2 = 6X$. The optimal cost is $J = \frac{2}{3}k^2 t_1 = (\frac{8}{3}k^3 X)^{1/2}$. The minimum value of x_2 occurs when $t = \frac{1}{2}t_1$ and is equal to $-\frac{1}{4}kt_1$, which is greater than or equal to -1 when $t_1 \le 4/k$ and so $X \le 8/3k$.

For these values of X the restriction on x_2 has no effect. When $X > 8/3k$, the optimal trajectory will include a portion of the boundary of the permitted region. Along $x_2 = -1$, $u_1 = 0$ and x_1 decreases linearly, from the state equations. If we suppose that the trajectory meets the line $x_2 = -1$ at $t = t_2$, then for $0 < t < t_2$,

$$z_2 = -k - At, \quad x_2 = -kt - \tfrac{1}{2}At^2, \quad x_1 = X - \tfrac{1}{2}kt^2 - \tfrac{1}{6}At^3.$$

From the continuity conditions at t_2, $z_2(t_2) = 0$ and also $x_2(t_2) = -1$, so that $t_2 = 2/k$ and $A = -\frac{1}{2}k^2$. With these values, we find that $x_1(t_2) = X - 4/3k$. For $t > t_2$, $x_1 = x_1(t_2) - (t - t_2)$, and, by symmetry, we should have $x_1 = \frac{1}{2}X$ at $t = \frac{1}{2}t_1$. Therefore, $t_1 = X + 4/3k$ and the optimal cost is $J = \frac{1}{2}k^2X + \frac{4}{3}k$.

10.8 The controls u_1 and u_2 are the velocity components and the maximum speed is equal to unity. To move between the two given locations in the shortest possible time is equivalent to making the distance travelled as short as possible. Since the maximum speed is unity, the minimum time and the minimum distance are equal in magnitude.

Forming the Hamiltonian in the usual way, we have the co-state equations $\dot{z}_1 = 0$, $\dot{z}_2 = 0$, and we must take the supremum of $z_1u_1 + z_2u_2$. Hence we choose u_1 and u_2 to lie on the boundary of the control set and we can write $u_1 = \cos\alpha$, $u_2 = \sin\alpha$, for some constant α. The optimal trajectory is therefore $x_1 = X_1 + t\cos\alpha$, $x_2 = X_2 + t\sin\alpha$. To reach the target, we must have $t_1\cos\alpha = Y_1 - X_1$, $t_1\sin\alpha = Y_2 - X_2$ and $t_1^2 = (Y_1 - X_1)^2 + (Y_2 - X_2)^2$.

With the restriction on the state variables included, the Hamiltonian takes the form $H = -1 + z_1u_1 + z_2u_2 + \lambda(1 - x_1^2 - x_2^2)$ and the co-state equations are $\dot{z}_1 = 2\lambda x_1$, $\dot{z}_2 = 2\lambda x_2$. The trajectory is made up of sections of two kinds: either $\lambda = 0$ and z_1 and z_2 are constant, or $x_1^2 + x_2^2 = 1$. Where the straight and circular portions of the trajectory intersect, the vector $\mathbf{z}$ must be parallel to the control vector $\mathbf{u}$, since we must maximize the scalar product $z_1u_1 + z_2u_2$. Since $\mathbf{z}$ is continuous at the intersection, $\mathbf{u}$ must also be continuous. But $\mathbf{u}$ is the velocity of the point as it moves along its path, so it follows that the straight portion of the trajectory must meet the bounding circle tangentially. The optimal trajectory is therefore given by

$$x_1 = -2 + t\cos\tfrac{1}{6}\pi, \quad x_2 = t\sin\tfrac{1}{6}\pi, \quad \text{for } 0 < t < \sqrt{3},$$

$$x_1 = \cos(\tfrac{2}{3}\pi + \sqrt{3} - t), \quad x_2 = \sin(\tfrac{2}{3}\pi + \sqrt{3} - t),$$
$$\text{for } \sqrt{3} < t < \sqrt{3} + \tfrac{1}{3}\pi,$$

$$x_1 = 2 - (t_1 - t)\cos\tfrac{1}{6}\pi, \quad x_2 = (t_1 - t)\sin\tfrac{1}{6}\pi,$$
$$\text{for } \sqrt{3} + \tfrac{1}{3}\pi < t < t_1,$$

where $t_1 = 2\sqrt{3} + \frac{1}{3}\pi$.

Exercises 11 (p. 197)

11.1 To maximize P we must minimize $-P$, so the Hamiltonian is given by

$$H = ux + kz(x - ux - c),$$

and the co-state equation is $\dot{z} = -u + k(1 - u)z$. Since u must lie between 0 and 1, we choose $u = 1$ when $z < 1/k$ and $u = 0$ when $z > 1/k$. Since the final state is not specified, we must have $z(T) = 0$ and we must end by using the control $u = 1$ and $z(t) = T - t$. This control must be applied for the whole time if $z(0) \leq 1/k$, that is, if $T \leq 1/k$. In this case, $\dot{x} = -kc$ and $x = X - kct$. Since $X > c$ the final value of x is positive and the optimal cost is given by

$$P = \int_0^T x \, dt = (X - \tfrac{1}{2}kcT)T.$$

When $T > 1/k$, we use $u = 0$ for $0 < t < T - 1/k$ and $u = 1$ for $T - 1/k < t < T$. Solving the state equation, we find that $x = c + (X - c)\exp(kt)$ for the first section, and $x = (X - c)\exp(kT - 1) + kc(T - t)$ for the second section. The optimal value of P is given by

$$P = (1/k)[(X - c)\exp(kT - 1) - \tfrac{1}{2}c].$$

A serious inadequacy of this model is that it assumes the farmer's ability to sow unlimited quantities of grain. Both land and labour in abundance must be available.

11.2 The optimal control is $u = h$ when $z < 1$ and $u = 0$ when $z > 1$. When $z = 1$, u is undetermined, but in this problem it is impossible for z to be equal to one except instantaneously. The terminal state is specified, and $x(0) = x(T) = 1$. The solution of the co-state equation shows that $z = 1$ at one value of t only. Suppose the switch occurs at $t = t_1$. If we set $u = h$ for $0 < t < t_1$, we find that $x = \exp[-(h - 1)t]$. For $t_1 < t < T$, we switch to $u = 0$ and find that $x = \exp(t - ht_1)$. The target is reached at time T provided $t_1 = T/h$. The catch is given by

$$-J = \int_0^{t_1} hx \, dt = \frac{h}{h-1}\left[1 - \exp\left(-\frac{h-1}{h}T\right)\right].$$

Another possibility is that we begin by using $u = 0$ and switch to $u = h$ at time t_2. In this case, $x(t_2) = \exp(t_2)$, and, for $t > t_2$, $x = \exp[ht_2 - (h - 1)t]$. To reach the target we must take $t_2 = (h - 1)T/h$ and the catch is

$$-J = \int_{t_2}^{T} hx \, dt = \frac{h}{h-1}\left[\exp\left(\frac{h-1}{h}T\right) - 1\right].$$

This second strategy yields the larger catch, and so is the optimal solution.

11.3 The controls that can be employed are $u = 0$ or $u = 2$ depending on whether $z > 1$ or $z < 1$. Exceptionally, we can use $u = 1$ when $z = 1$ and $x = 1$. At time T we must have $z(T) = 0$ provided $x(T) > 0$. There are three possible strategies.

In the first, we use $u = 2$ throughout. Then $\dot{x} = -x^2$ and $\dot{z} = -2 + 2xz$, with solutions

$$x = \frac{1}{t + c}, \quad z = \frac{2(t + c)(T - t)}{T + c}.$$

At $t = 0$, $x = \varepsilon$, so $c = 1/\varepsilon$ and the catch is $-J = 2\ln(1 + \varepsilon T)$. This solution is valid provided $z(t) \leq 1$, so that we must have $T \leq (2 - \varepsilon)^{-1}$. The second possibility is that we use $u = 0$ for $0 < t < t_1$ and $u = 2$ for $t_1 < t < T$, with $z(t) = 1$. The value of $z(t_1)$ can be found from the expression above, so that $2(t_1 + c)(T - t_1) = T + c$. Also, with $u = 0$, the state variable is given by $x = 1 + \tanh(t - \alpha)$, where $\tanh \alpha = 1 - \varepsilon$. Equating the two values of x at $t = t_1$ gives the equation $1 + \tanh(t_1 - \alpha) = (t_1 + c)^{-1}$. We thus have two equations to determine the values of c and t_1. The catch is given by

$$-J = 2\ln\left(\frac{T + c}{t_1 + c}\right).$$

This solution is valid provided $x(t_1) < 1$, that is, $t_1 < \alpha$. If we continue with $u = 0$ past $x = 1$, z begins to increase and will remain greater than unity and no switch will be possible. The third possibility is that we use $u = 0$ for $0 < t < \alpha$ until $x = 1$. For $\alpha < t < T - 1$, we use $u = 1$, with $x = 1$ and $z = 1$. For $T - 1 < t < T$ we use $u = 2$. This is a possible solution, provided $T \geq 1 + \alpha$ and the optimal catch is given by $-J = T - 1 - \alpha + \ln 4$.

For T sufficiently large, this optimal solution means that fishing only commences when the stock has reached the size $x = 1$, when fishing takes place at a rate that keeps the level of the stock unchanged. There is a small additional terminal gain achieved by fishing out the stock completely.

11.4 The final state is unspecified so $z(T) = 0$. The Hamiltonian and the co-state equation are given by $H = u + z(2x - x^2 - u)$ and $\dot{z} = -z(2 - 2x)$. To find the supremum of H we must choose $u = h$ when $z < 1$ and $u = 0$ when $z > 1$. When $z = 1$, the value of u is unspecified, but this can only occur for a finite time interval if $\dot{z} = 0$, that is, if $x = 1$. Since $z(T) = 0$ we must end by using $u = h$ and the solution of the co-state equation shows that $z(t) = 0$, unless we switch to the execptional control when $x = 1$.

Suppose first that $h < 1$, and write $h = 1 - p^2$. Then the solution of the state equation is $x(t) = 1 + p\coth(pt + \alpha)$, where $\tanh\alpha = p$. At the terminal time, $x(T) > 1$ and the optimal catch is given by $J = \int_0^T u\,dt = hT$.

Now suppose that $h > 1$ and write $h = p^2 - 1$. With $z = 0$ again, and $u = h$, the solution is $x(t) = 1 + p\cot(pt + \alpha)$, where $\tan\alpha = p$. Since $x(T) \geq 0$, this solution is only valid when $T \leq (\pi - 2\alpha)/p$. The optimal catch is again equal to hT.

The third case occurs when $h = p^2 + 1$ and $T > (\pi - 2\alpha)/p$. We use the control $u = h$ for $0 < t < t_2$, with $x(t_2) = 1$ so that $t_2 = (\pi - 2\alpha)/2p$. Then we continue with $z = 1$ and with $x = 1$. The control is not determined by the supremum of H in this case, but the state equation shows that, since x is constant, we must have $u = 1$. Continue with this control until $t = t_3$, say. Then for $t_3 < t < T$ we must have $u = h$ again, and the trajectory is $x(t) = 1 + p\cot(pt - pt_3 + \frac{1}{2}\pi)$. With $x(T) = 0$, this equation gives $T - t_3 = (\pi - 2\alpha)/2p$. The optimal catch is given by

$$J = ht_2 + (t_3 - t_2) + h(T - t_3) = T + p(\pi - 2\tan^{-1} p).$$

11.5 If we put $x_1 = y$, $x_2 = dy/dt$, the state equations are $\dot{x}_1 = x_2$ and $\dot{x}_2 = -\varepsilon^{-1}x_2 + u$. From the Hamiltonian, we find that the control $u = z_2$ and that $\dot{z}_1 = 0$ and $\dot{z}_2 = -z_1 + \varepsilon^{-1}z_2$. At $t = 0$, $x_1 = 1$, $x_2 = 0$, and $z_2 = -c$. The last condition comes from setting $H = 0$ and the obvious requirement that we start with a negative value of u. At $t = T$, $x_1 = 0$, $x_2 = 0$, and $z_2 = c$.

When t is not close to either 0 or T, the solutions of the state and co-state equations have the form $z_2 = -A$, $z_1 = -\varepsilon^{-1}A$, $x_2 = -\varepsilon A$, $x_1 = B - \varepsilon At$, where A and B are constants. Near $t = 0$, we put $t = \varepsilon\tau$ and the equations now have the forms

$$\frac{dx_1}{d\tau} = \varepsilon x_2, \quad \frac{dx_2}{d\tau} = -x_2 + \varepsilon z_2, \quad \frac{dz_1}{d\tau} = 0, \quad \frac{dz_2}{d\tau} = z_2 - \varepsilon z_1.$$

The solutions of the co-state equations that satisfy the conditions at $t = 0$ are given by

$$z_1 = -\varepsilon^{-1}A, \quad z_2 = -A + (A - c)e^{\tau}$$

and the matching with the solution found when t is not small gives $A = c$. The corresponding solutions of the state equations are given by

$$x_2 = -\varepsilon c + \varepsilon c\,e^{-\tau}, \quad x_1 = 1 - \varepsilon^2 c\tau + \varepsilon^2 c[1 - e^{-\tau}],$$

and, for large τ, $x_2 \to -\varepsilon c$, $x_1 \sim 1 + \varepsilon^2 c - \varepsilon ct$, which match with the solution when t is not small if $B = 1 + \varepsilon^2 c$. When t is close to T, we put $t = T - \varepsilon\theta$. The equations are now given by

$$\frac{dx_1}{d\theta} = -\varepsilon x_2, \quad \frac{dx_2}{d\theta} = x_2 - \varepsilon z_2, \quad \frac{dz_1}{d\theta} = 0, \quad \frac{dz_2}{d\theta} = -z_2 + \varepsilon z_1.$$

The solutions of the co-state equations that match with the interior solution and satisfy the conditions at $t = T$ are

$$z_1 = -\varepsilon^{-1}c, \quad z_2 = -c + 2c\,e^{-\theta}.$$

From the state equations and the conditions on the state variables, we find that

$$x_2 = -\varepsilon c + \varepsilon c\,e^{-\theta}, \quad x_1 = \varepsilon^2 c\theta + \varepsilon^2 c[e^{-\theta} - 1].$$

Hence we find that, as $\theta \to \infty$. $x_2 \to -\varepsilon c$ and $x_1 \sim \varepsilon c(T - t) - \varepsilon^2 c$. The interior solution for x_1 has been found to be $x_1 = 1 + \varepsilon^2 c - \varepsilon ct$, and the value of T is found by matching these two expressions for x_1. Hence,

$$T = \frac{1}{\varepsilon c} + 2\varepsilon,$$

with exponentially small terms omitted. The optimal cost is given by

$$J = \int_0^T \tfrac{1}{2}(c^2 + z_2^2)\,dt = \tfrac{1}{2}c^2 T + \int_0^{\varepsilon^{-1}T} \tfrac{1}{2}\varepsilon c^2[2e^{-\theta} - 1]^2\,d\theta$$
$$= c^2 T - \varepsilon c^2,$$

again neglecting exponentially small terms, so that

$$J = \frac{c}{\varepsilon} + \varepsilon c^2.$$

11.6 When $X_2 \geq t_s$, the strategy that was used when shortselling was permitted can be applied, since x_2 now remains non-negative throughout. Thus, $u = 1$ when $0 < t < t_s$ and $u = -1$ when $t_s < t < T$. The only difference is that a term $-sX_2$ must be added to the previous values of $\dot{x}_1$, so that the final value of $x_1(T)$ is reduced by sX_2T. Hence the maximum profit is

$$P = \frac{p_1^2 - p_1 s + 2s^2}{4(p_1 + s)}\,T^2 - sX_2T.$$

With $p_1 = 2$ and $s = 1$, $P = \frac{1}{3}T^2 - X_2T$, provided $X_2 \geq \frac{1}{3}T$. In this case, the transactions must lead to a loss. The decrease in the value of P is due to the extra storage charge for the increased stock.

When $0 < X_2 < t_s$, the strategy is to set $u = 1$ until the stock has been reduced to zero, $u = 0$ until the time t_s is reached, and then $u = -1$ for $t_s < t < T$, as before. With $X_2 = 0$, the first stage was not present, and the profit was found in (11.79). Repeating the calculation with $X_2 > 0$, we find that

$$P = \frac{p_1(p_1 - s)}{4(p_1 + s)} T^2 - \tfrac{1}{2}(p_1 + s)X_2^2.$$

With $p_1 = 2$ and $s = 1$, $P = \frac{1}{6}T^2 - \frac{3}{2}X_2^2$, which is positive since $X_2 < \frac{1}{3}T$.

For general price structures, the geometrical method can be used as before. The only change is in the determination of the discontinuities in z_2 and the position of the periods of inactivity during which x_2 is equal to zero.

11.7 The extra term changes the first co-state equation to $\dot{z}_1 = -rz_1$, so that $z_1 = \exp[r(T - t)]$ and then the second co-state variable is given by

$$z_2 = p(T) + (s/r)\{1 - \exp[r(T - t)]\}.$$

The optimal control is

$$u = \operatorname{sgn}\Big(p(t) - p(T) - (s/r)\{1 - \exp[r(T - t)]\} \Big).$$

The same geometrical method for finding the solution can be used, but the straight lines for the z_2 curves must be replaced by exponential curves.

If the particular values for $p(t)$ and the constants are chosen, then the switching time t_s is the root of $\exp(T - t_s) = 1 + 2t_s$. With shortselling prohibited, we use $u = 0$ for $0 < t < t_s$ and $u = -1$ for $t > t_s$. The solution of the equation for x_2 has the same form as before, but the equation for x_1 is now $\dot{x}_1 = rx_1$ for $0 < t < t_s$ and $\dot{x}_1 = -p(t) + x_1 - x_2$ for $t > t_s$. Solving this equation we can determine x_1 and hence find the profit.

When $T < 2.513$, $t_s < \frac{1}{2}T$ and the profit is given by

$$P = X_1[e^T - 1] + p_0(T - 3t_s) + 4 + T + 3t_s + 4t_s^2 - 4\exp(\tfrac{1}{2}T).$$

When $T > 2.513$, $t_s > \frac{1}{2}T$ and we do not start buying until after the price has passed its minimum value at $t = \frac{1}{2}T$. The solution in this case is

$$P = X_1[e^T - 1] + p_0(T - 3t_s) + 3T - 9t_s + 4Tt_s - 4t_s^2.$$

Exercises 12 (p. 212)

12.1 The optimal control with the parameters given in the text decreases to a shallow minimum, then increases to a maximum slightly lower than its initial value. It then drops sharply and becomes negative shortly before the end of the cycle.

With $c_1 = 0$, the initial dip in the control does not appear, but otherwise the behaviour is quite similar. The maximum value is about 20 per cent higher than in the test solution.

If $c_2 = 0$, the equation for x_2 is $\dot{x}_2 = -c_1 x_2$ and it is impossible to satisfy the two conditions $x_2^0 = 0$ and $x_2^1 = 0$ except by $x_2(t) = 0$, in which case there is no flow of blood.

With $c_3 = 0$, the control begins at a level about 30 per cent higher than in the test case but then decreases to its final negative values monotonically.

With $c_4 = 0$, the initial value is up by about 40 per cent, but it rapidly decreases to values comparable with the test case. A minimum is reached, followed by a slight maximum, and then the rapid decrease to negative values is again obtained.

With $c_5 = 0$, the initially high value of the control, comparable to that in the previous case, is followed by a monotonic decrease.

It follows that the variation of the control pressure in the cycle is only matched in these simplified models when c_4 is put equal to zero. Even then, the size of the control is considerably increased.

12.2 Good luck!

Index